Spiral Structure in Galaxies

Spiral Structure in Galaxies

A Density Wave Theory

G. Bertin and C. C. Lin

The MIT Press
Cambridge, Massachusetts
London, England

This book was set in Palatino by Windfall Software using ZzTEX and was printed and bound in the United States of America.

Library of Congress Cataloging-in-Publication Data

Bertin, G. (Giuseppe)
 Spiral structure in galaxies : a density wave theory / G. Bertin, C. C. Lin
 p. cm.
 Includes bibliographical references and index.
 ISBN 0-262-02396-2
 1. Spiral galaxies. 2. Density wave theory. I. Lin, C. C. (Chia-Ch'iao), 1916– II. Title
QB858.42.B47   1995
523.1'12—dc20

95-11635
CIP

Contents

Preface

In this book we present a theory of spiral structure in galaxies. Our primary goal is to make the key concepts (the basic physical picture and the relevant astrophysical implications) explicit, and therefore available for criticism and accessible to the widest possible audience. The theory has matured considerably since its initial conception in the 1960s, but it still leaves a number of very important issues unresolved. This "incompleteness" is welcome, though, because a field with unresolved issues is an open field, able to attract more interest than a field where all the answers have already been given. We deliberately downplay history in the book, mentioning only briefly many of the observational tests of the 1960s and 1970s that played a crucial role in the early development of the density wave theory. Our focus is on the growth of the theory, in which the emphasis has now considerably changed toward recognizing the importance of a proper modeling of the basic state of galaxy disks.

The book aims at describing a coherent framework for the problem of spiral structure in galaxies and is not written in the spirit of a review article. After almost seventy years since the discovery of the differential rotation in our galaxy and after thirty years of extensive work on density waves, it is clear that collecting all the relevant references and credits would be a formidable task, of interest to different readers from those we have in mind. Initially, we had even thought to cite only general references and review papers, except, of course, for the credits required by the figures selected. In dealing with the second level of part III, we have slightly departed from this initial plan, considering that some interested readers might want to check the details and the derivations in the original papers. This explains why the reference lists are incomplete. Many more papers and contributions are reviewed in the references that we do list, but we apologize in advance to everyone who finds that some of their relevant articles are not quoted explicitly.

The book is addressed to a wide audience and is as nontechnical as we could make it. It is written at two levels. Note that the illustrative material for the first two-thirds of the book is mostly photographic, while that for the last third consists mainly of diagrams. At the first level (essentially, part I and part II), the text is mostly descriptive and "physical," aimed at general readers curious in astronomy, with some scientific background (anyone who has taken science undergraduate courses should be able to get interested and to follow). At the second level (essentially part III and, to a large extent, chapter 0, which is a kind of extended summary), the book becomes more explicit on dynamical mechanisms and slightly more mathematical. At this second level, the book is aimed at graduate students and scientists actively involved in either astronomy or related subjects (such as plasma physics or meteorology), allowing them to see the framework of the theory reduced to its backbone, with reference to the relevant research papers but with no details or derivations. For this purpose, some of the concepts described earlier in physical terms are revisited in more quantitative form, with the aid of a few selected diagrams. This process involves a small amount of repetition, which may be useful to readers not familiar with the subject. At the second level of part III, we have avoided the use of mathematics as much as possible, especially because we believe that the subject has matured to a stage where the main results and concepts can be described in simple physical terms, with the use of a few diagrams. Still, we think that insisting at all costs on a formula-free, superficially popular kind of presentation would have made the text unnecessarily obscure or ambiguous, and we have tried to clear away such ambiguity through explicit use of proper scientific terms and a few mathematical definitions.

0   Introduction

0.1   Morphology and Semiempirical Approach

Spiral structures in galaxies occur on many scales. In this monograph we shall address the problem of spiral structures on all scales, but with specific emphasis on global patterns (i.e., structures on large scales, of a few kpc for a galaxy such as the Milky Way). The Hubble morphological classification scheme [4] empirically correlates the pitch angle of spiral structure with other physical characteristics of galaxies, such as bulge size and gas content, which are not expected to change significantly in one typical revolution time. In the words of Jeans [5], "The great nebulae exhibit an enormous difference of structural detail, but Hubble, who has devoted much skill and care to their classification, finds that most of the observed forms can be reduced to law and order." Therefore, the primary emphasis of this study will be on the *intrinsic*, slowly evolving characteristics of disk galaxies. Our approach will be semiempirical, that is, our dynamical studies will always attempt to address issues directly raised by empirical facts and will always be guided by the various clues offered by observations. We shall avoid too much speculation on dynamical mechanisms and physical scenarios that do not have sufficient empirical basis. In particular, we shall focus on spiral galaxies with quite regular global structures (see figure 0.1), commonly called "grand design" galaxies because they pose challenging problems in the clearest form. A detailed discussion of the various observed morphologies will be given in part I and part II, especially chapters 1, 3, 4, and 5.

Figure 0.1
M81 [11].

0.2 Coexistence of Small- and Large-Scale Structures

Even the best examples of grand design galaxies, when inspected on
the small scale, display a number of local features, such as "feathers,"
"spurs," "broken arms," or "clumps." These structures appear to be
related to the dynamics of the gaseous component (see figure 0.2). In
some disk galaxies (see figure 0.3) a grand design is absent and only
flocculent spiral features are observed. In the following chapters (especially chapters 1 and 3) we shall elaborate further on these points. Here
we want to emphasize the coexistence of small- and large-scale structures from the very beginning, in order to give a proper statement of
the goals of the present monograph.

The first major goal of our study is to explain how spiral structures
are excited and maintained, starting from the large scale; the same general principles should also be able to explain the smaller-scale structures. In addition, the same theory should explain the absence of global
structures in many galaxies as well. This point is made clearly by the
following statement of the problem as given by Oort [8]:

Figure 0.2
NGC 618 [11].

In systems with strong differential rotation, such as is found in all non-barred spirals, spiral features are quite natural. Every structural irregularity is likely to be drawn out into a part of a spiral. But *this* is not the phenomenon we must consider. We must consider a spiral structure extending over the whole galaxy, from the nucleus to its outermost part, and consisting of two arms starting from diametrically opposite points. Although this structure is often hopelessly irregular and broken up, the general form of the large-scale phenomenon can be recognized in many nebulae. . . . It may be practical, at least in the present stage of the knowledge, to separate the problem into two parts: (a) How did spiral structure originate? (b) How does it persist once it has originated?

Figure 0.3
NGC 2841 [11].

The second major goal is to develop a dynamical classification of galaxies, in support of the empirical morphological classification; that is, we wish to set up (e.g., with the help of an atlas of dynamical models) a general correspondence between observed morphologies and intrinsic dynamical characteristics for the categories of galaxies under investigation. The third major goal is to develop the method for a proper dynamical modeling of the basic state of individual galaxies and for the characterization of their structural parameters based on the constraints set by the observed spiral structure. As a result, the concerns of this monograph are strongly related to other astrophysical issues, such as the problem of the amount and distribution of dark matter (especially on the large scale) and the processes of star formation (on the small scale).

0.3 Excitation of Spiral Structure

Probably the first specific questions that come to mind while looking at the beautiful spiral arms in disk galaxies are these: What is the en-

Figure 0.4
M51 [11].

ergy source for such structures? How are they excited? Without reviewing all the mechanisms that have been proposed, we can say that the
empirical evidence argues for at least three options: (1) Spiral structure is generated by external excitation: a companion or an encounter
is involved in the process (see figure 0.4). (2) A well-formed oval/bar
drives the spiral structure (see figure 0.5; but then we should ask how
the bar was formed). (3) The intrinsic dynamics of the disk may naturally give rise to the morphologies observed, even in the absence of
bars and tidal interaction. Other proposed mechanisms, such as continuous gas infall from the extragalactic environment or "propagating
star formation" via a sequence of supernova explosions, are not empirically favored (although they may play a role in the overall dynamics of
disk galaxies), because they do not address the issue of the formation of
smooth large-scale stellar arms that are often observed (see figure 0.6).
We also note that driving by bars or tidal interaction requires well-
defined intrinsic characteristics of the disk, which should be ready to
"accept" (i.e., respond favorably to) the driving.

Figure 0.5
NGC 1398 [11].

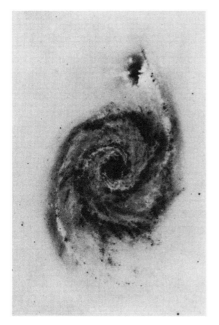 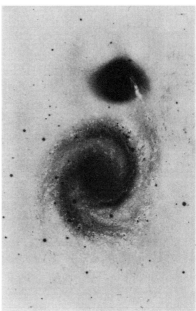

Figure 0.6
M51 through a red filter [12].

These three possibilities are not necessarily mutually exclusive. Thus it would be desirable to study in detail the mechanisms required by each of the three possible scenarios and make sure that the invoked basic states are not ad hoc but physically plausible. It turns out that it is very difficult to determine, directly from the observational data, which of the above scenarios is significant for given *individual* objects. For example, even in those cases where some kind of interaction through a galactic encounter is obviously at work (see figure 0.4), it remains to be proven how much of the observed large-scale structure is actually due to the presence of the "companion" and how much of it instead essentially preexisted the encounter between the two galaxies.

In this monograph we prefer to focus on the last interpretation, namely, that the global spiral structure in most galaxies is generally associated with intrinsic, natural "modes" or long-lasting but slowly evolving wave patterns, dictated by the dynamics of the disk. This is most likely to work as a general explanation because spiral arms are not always bisymmetric and galactic encounters are too infrequent to account for the prevalence of spiral structures. The standing wave pattern in question could be small or finite in amplitude. Still, the exact fraction of objects for which such a modal approach (see section 0.4 and chapter 4) is justified is hard to quantify. In the following chapters we shall stress how natural the modal approach is by bringing out all the astrophysical implications of this point of view. Focusing on the intrinsic dynamics of the disk is further desired because, even for those cases where spiral structure should owe its existence to external driving, the excited structure may well be close to a single damped mode determined by the intrinsic characteristics of the disk.

The possibility that spiral structure may be self-excited in the disk is quite natural. There is plenty of energy stored in the differential rotation of the disk, which can be released for the formation of spiral arms. Indeed, shear flows in ordinary hydrodynamics, in addition to being subject to transient instabilities with no net amplification [9], are known to be unstable whenever a mechanism can trigger the transfer of momentum (or angular momentum in rotating flows) across the region where the waves comove with the material fluid particles of the basic state (see Prandtl [10], Heisenberg [3], and the extensive discussion given by Lin [6] and by Drazin and Reid [2]). The detailed mechanisms can be subtle, but the simple result is that the available energy present in the form of shear motions can indeed be released.[1] In light galaxy disks, this process is favored by the presence of the cold gas component. Indeed, detailed dynamical studies (see chapter 4) show that galaxies with a relatively cool stellar disk can support self-excited global spiral structures, generally with two arms, through further gravitational coupling with a significant amount of the very cool gas in the outer disk. (A feedback process from the central regions of the galaxy is also essential for the formation of modes; cf. sections 0.4 and 0.7.) On the other hand, for warmer stellar disks, the spiral activity develops mostly in the cold gas component. Heavier disks are prone to bars and these, in turn, drive their own spiral arms. Tidal interactions may sometimes give temporarily more coherence to an otherwise multiple-armed spiral galaxy; however, they may often "damage" the disk by producing heating and thickening of the stellar component.

0.4 Three "Persistence" Problems and Maintenance of Spiral Structure

Especially because the appearance of spiral structure correlates with a number of physical properties that are likely to change slowly in

---

1. Paradoxically, usually with the help of dissipative mechanisms.

time, the most appealing procedure for astrophysical applications is to develop a semiempirical approach starting with the hypothesis of quasi-stationary spiral structure (QSSS). As is well known in hydrodynamics, even if local features may be rapidly evolving, large-scale, global patterns are often slowly evolving or quasi-stationary (the prefix "quasi" should be stressed—see the detailed discussion in chapters 2 and 4). Let us then assume that spiral structures can be long-lasting, as a working hypothesis. Note that this possibility was clearly recognized by B. Lindblad [7] in his paper entitled "On the possibility of quasi-stationary spiral structure in galaxies," despite indications from the computational models studied by his son Per Olof that the spiral structure might be "quasi-periodic" or "regenerative." We should then consider the dynamical conditions under which QSSS can be expected to hold and compare these conditions with observed conditions in the astrophysical context.

Winding Dilemma An immediate persistence problem that has to be faced is the so-called winding dilemma. If spiral arms were material arms, they would quickly wrap up because of differential rotation and their shape would change significantly over one typical period of revolution. A solution to this dilemma is obtained if the arms result from density waves (see chapter 2). One advantage of the QSSS hypothesis is that it is very simple and can be easily quantified in the form of predictions to be tested. For many astrophysical applications, such as shocks and star formation processes, the relevant timescales are short enough, so that we can work *as if* the spiral structure were steady, even when it is evolving on a timescale fairly close to the dynamical time scale. On this basis, a large set of important calculations and successful observational tests of the "density wave theory" were performed in the 1960s and in the early 1970s so as to establish consistency with observational data (see chapters 5 and 6). For this purpose, the mathematical basis was a "local" theory for the determination of the pattern, the so-called dispersion relation for tightly wound spirals (see chapters 8 and 9), where spiral arms are regarded as the manifestation of a single wave, with a pattern frequency as a parameter to be determined from a best fit to the available observational data, and with amplitude distribution along the spiral arm to be empirically specified.

Waves and Modes Here a second persistence problem naturally arises because, in the highly inhomogeneous and dispersive environment of galaxy disks, density waves ("wave packets") are generally expected to evolve and to propagate (see chapter 9). However, if a global-scale spiral structure is under consideration, boundary conditions and feedback processes play important roles in the dynamics of the underlying density waves. Thus it is quite natural to imagine that when we look at a beautiful grand design spiral, such as M81, NGC 5364, NGC 1300, or even M51, we are indeed looking at a global pattern produced by internal discrete global modes of the disk. These modes are composed of waves that may satisfy "locally" a dispersion relation as argued above. The resulting pattern, while dynamically supported as a self-sustained global mode, may in turn be well approximated by one branch of such waves satisfying the local dispersion relation. In order for the QSSS hypothesis to apply, the large-scale spiral structure must be dominated by one global mode or by a small number of such modes. Once the modal approach is accepted, mastering the relation between the properties of the basic states and the properties of the relevant spiral modes and identifying the realistic basic states can lead to a unified view of morphologies and of astrophysical processes in spiral galaxies. In the development of the studies of spiral structure in galaxies, the major shift of emphasis from wave theory to modal theory occurred only in the mid-1970s, even though the modal picture was necessary for the QSSS hypothesis to hold and the concept of the feedback process was introduced in the late 1960s. Now the modal theory has become well developed to cover most of the key morphologies that are found in the Hubble classification scheme (see chapters 4 and 10).

Heating of the Stellar Disk   A purely stellar disk would, in general, be subject to a perennial conversion of ordered kinetic energy into random motions as a result of collective instabilities and possibly of the scattering of orbits by "external" driving forces. Thus it seems that a purely stellar disk would be subject to heating and, once hot and stable enough, could not return back to a cooler state. Therefore, we would expect that spiral structure, even if present initially in the form of self-excited global modes, would rapidly die out. The solution to this third persistence problem is provided by the presence of the cold gaseous component. We find that (see chapters 3 and 4), at least for normal (nonbarred) spirals, the cold and dissipative gas can keep the outer disk dynamically cool, as a "thermostat," via a process of self-regulation. In the absence of gas, at least for normal spiral galaxies, spiral structure could not be long-lasting. We shall see that this point plays a crucial role in the identification of the appropriate basic states, that is, in the proper modeling procedure (see chapter 7).

0.5   Roles of the Cold, Dissipative Gas Component

The cold interstellar medium plays a dominant role in determining the observed spiral structure in galaxies not only because it is the site of the generation of young and bright objects, but also because it is both a source of excitation (Jeans instability) and of regulation (because it is dissipative) for the overall dynamics of the disk. Regulation is present at two different levels. On the one hand, the cold gas component, through shocks, can provide a saturation mechanism for growing (self-excited) spiral modes at finite amplitudes. On the other hand, if the stellar random motions are not too large, so that the dynamics of the stars is sufficiently well coupled gravitationally to the dynamics of the cold gas, some cooling in the interstellar medium can compensate for the heating that may occur in the stellar component. As a result, the disk as a whole can be maintained on the margin of local dynamical stability; in turn, such a state of marginal stability with respect to local Jeans collapse allows for the excitation of large-scale spiral modes. This latter mechanism is a process of self-regulation.

0.6   The Dynamical Window

Studies in galactic dynamics fall into various categories, such as the identification of self-consistent equilibrium solutions, the analysis of stellar orbits in a specified potential, the linear stability analysis of a given model or set of models, or the study of the evolution of a dynamical system (often carried out by means of n-body simulations).

Obviously, the properties that are derived from these studies are characteristic of the models chosen. In general, different models are found to possess very different dynamical behavior. In addition, this behavior is sensitive to some ingredients, such as the presence of a small amount of cold gas or the finite thickness of the stellar disk, that at first sight might appear to be of secondary importance. Therefore, the process of applying dynamical studies to real galaxies hinges on the crucial step of setting up an appropriate model for the galaxy or for the class of galaxies that one wishes to describe. Actually, observations provide just a partial, blurred view of the structure of galaxies.

Not only are key structural parameters, such as disk mass, still uncertain by a factor of 2, but other quantities, such as the radial velocity dispersion profile in the stellar disk of external galaxies, are likely to be well beyond the reach of direct measurement for some time to come. One important goal of our studies of the dynamics of spiral structure is to take advantage of the sensitivity of collective dynamics to the basic structural parameters and to bring the basic properties of spiral galaxies into sharper focus by looking through the "dynamical window." In other words, we are aiming at the identification of the relevant models of basic states, namely, models that are dynamically self-consistent and, in addition, compatible with the main observed large-scale spiral phenomena. Therefore, this line of research does depend heavily on the observational windows for its empirical foundation, and yet it can provide independent additional information on the structure of galaxies (see chapter 7). As a result of this general perception, observational studies related to the density wave theory (see chapters 5 and 6) are made in two stages. At an earlier stage, one basically looks for evidence of waves and modes; at a later, advanced stage, the application of waves and modes to spiral galaxies probes in further detail the intrinsic structure of their basic state.

0.7 Prevalence of Two-Armed Grand Designs

Another natural question is why grand design spiral structure is usually bisymmetric (there are also examples where the regular global structure has significant one-armed or three-armed components). It should be noted that, as first pointed out by Zwicky [12], the global spiral structure appears more regular and smoother in red images (see right frame of figure 0.6) than in the blue. This indicates (see section 0.9) that the large-scale waves are likely to be primarily carried by the stellar disk in these systems, while less regular structure is gas-dominated (see figure 0.4). This conclusion is now strengthened by the beautiful images of spiral galaxies that are becoming available in the infrared K-band (around $2\,\mu$), which gives a direct view of the underlying old disk component. Large-scale structure requires feedback from the central regions of the disk in order to be maintained as a global mode (mesoscale structures may then form in the global spiral field). For the stellar component, the feedback is generally available for one- and two-armed perturbations (two-armed disturbances would then be more easily excited because of their greater efficiency in carrying angular momentum outwards), while perturbations with three or more arms have the feedback easily prevented by absorption at an orbital resonance of the stars with the rotating global spiral field, known as "inner Lindblad resonance" (see chapters 4 and 10). The precise statement on the occurrence of feedback depends on many factors, such as the presence and radial extent of the central bulge (see figure 0.1), the mass distribution in the disk, the velocity dispersion, but especially on the shape of the rotation curve (i.e., the profile of the rotation velocity of the disk as a function of galactocentric radius). The gas component has a different behavior at the resonance and responds on a smaller scale, so that it displays more easily multiple-armed structure (see figure 0.2). The extreme case is that of flocculent galaxies (see figure 0.3), where the density perturbations cannot be organized into a large-scale mode, presumably because the stars have excessive random motions and cannot participate in such a process.

0.8   Barred Spirals

So far we have mostly focused our attention on the subtle issues raised by the spiral structure in normal spiral galaxies. In contrast, barred galaxies can be understood in simpler terms from the physical point of view (although their mathematical description is quite complex, given that "local" analyses are hard to justify in this case). Analogies may be drawn with the theory of classical ellipsoids [1]. Barred galaxies are expected to be associated with relatively heavy disks, that is, with systems where the halo mass inside the optical radius is small. Their stellar disks should be quite warm because in cooler disks rapid heating would take place as a result of local gravitational instability. For these galaxies, the gas component is mostly passive with respect to the large-scale, open bar driving, dominated by the stellar component; of course, gas is still active in determining small-scale spiral structure. This latter morphology disappears when gas is absent, as in SB0 galaxies (see figure 0.7). Here it may be interesting to recall the description of NGC 2859 given by Sandage [11] in his Hubble Atlas: "This is the type example of the SB0$_2$. The bar is fuzzy and indistinct. There is a brightening of the rim of the lens at two opposite ends. There are no sharp boundaries anywhere." Earlier in the same atlas, he had stated: "SB0$_2$ galaxies have a nucleus, a lens, and a bar. The bar is *not* continuous from the nucleus to the rim of the lens. It consists of the central

Figure 0.7
NGC 2859 [11].

nucleus plus two regions of enhanced luminosity at diametrically opposite points of the periphery of the lens." This is precisely the picture
obtained from the modal approach, which turns out to be very useful
also for the interpretation of barred galaxies (see chapter 4).

0.9 Scales of Motion

An easily noted quantitative feature distinguishes stars from cold gas
material in the disk. A typical turbulent speed associated with the cold
gas component, that is, a typical random velocity associated with gas
clouds, is 6 km/sec. This number seems to be fairly constant, both
within a galaxy and from galaxy to galaxy. In contrast, for stars in
the main body of the disk, a typical velocity dispersion would be six
times larger (cf. the peculiar stellar motions in the solar vicinity). Thus,

while a star may wander around an annulus 2 kpc thick, gas clouds are constrained to move in thin annuli ($\Delta r \sim 300$ pc). This point is at the basis of the coexistence of large-scale structures and local features in spiral galaxies.

Another point to keep in mind is that spiral features noted in photographic images of galaxies are generally traced by bright and short-lived stars. If these stars are born with a speed relative to that of the pattern on the order of 30 km/sec, they are expected to travel in their lifetime a distance of 500 pc or less (in the frame of reference where the pattern is quasi-stationary). This scale should be kept in mind when applying the results of the density wave theory to intrinsically small objects (such as UGC 2259) or to intrinsically large objects (such as UGC 2885). It is not surprising that spiral structure is generally less well delineated in small galaxies than in large objects.

0.10 Viability of the Modal Approach for Normal Spirals

In contrast with the modal approach, one might consider dynamical scenarios where evolution takes place on the dynamical timescale of the epicyclic frequency of stellar orbits (in principle, rapidly evolving structures may be described in terms of the superposition of many modes; see chapter 11). As a criticism to the hypothesis of quasi-stationary spiral structure, three challenges to the modal approach have often been raised: (1) The observed spiral structures in galaxies are not as regular in appearance as the modes calculated. (2) The size of the observed spiral pattern is much larger than the (exponential) length scale of the distribution of stellar mass, often by a factor of as large as 3. (3) The observed stellar velocity dispersion in our galaxy seems to indicate that the galactic disk might be too stable for normal spiral modes to be self-excited; in addition, the stellar velocity dispersion is expected to increase with time.

As already suggested in sections 0.4 and 0.5, we argue that these challenges are easily resolved when one properly considers the role played by the cold interstellar medium. On the other hand, especially in relation to the last two of the issues raised above, it might appear somewhat surprising that the gaseous component can play an effective role, because the *total* amount of gas present in spiral galaxies is known to be small. In fact, even gas-rich (Sc) spirals have less than 10% of their mass in the form of gas. Here we should recall that these numbers usually refer to the ratio of the total mass in atomic hydrogen to an estimate of the total mass based on the properties of the observed rotation curve. However, if one refers to the *local* amount of gas (surface density) relative to the local density of the stellar disk $\alpha = \sigma_g / \sigma_*$, it is easily recognized that gas can be quite important in the outer parts of the galaxy disk.[2] In the radial range 3–4h_* (with h_* being the length scale of the exponential decay of the stellar density profile σ_*), α can easily exceed the value of 30%. It is in this radial range that we expect normal spiral structure to be excited. Thus the length scale for the mass distribution dynamically active is generally larger than h_*. The viability of the modal approach depends on a quantitative demonstration that these clues indeed lead to realistic scenarios for the structure of spiral galaxies.

0.11   A Physical Perspective

The ultimate purpose of theoretical studies is to discover and to develop models that will account for observational facts. It is thus important that we should always place strong emphasis on the empirical side. The crucial step is to determine, within the variety of dynamical scenarios that may be imagined, which dynamical models would actually be applicable to the physical systems, that is, to ascertain which physical situations would actually prevail and how often each scenario occurs in the real universe. This determination should obviously be based on comparison with observational data.

After reviewing the existing literature on global structure of spiral galaxies, both theoretical and observational, it is our conclusion that most of the observed global structures in the stellar system are primarily long-lasting and slowly evolving wave patterns maintained basically by intrinsic mechanisms. The modal approach is thus preferred. Therefore, the initiation of density waves and the formation of global stuctures may generally be traced to primordial times. In some cases, tidal interaction among galaxies may have played a role in the generation process. However, it should be kept in mind that such interactions (even for cases like that shown in figure 0.4) may often involve galaxies with *preexisting* global spiral structures. For gravitationally bound systems, continual tidal forcing is expected to play a role in the generation

---

2. The presence of other species of gas also helps. For dynamical considerations, the finite thickness of the stellar disk also contributes to an increase of the effective value of α, and of the length scale of the "active mass" density distribution (see chapter 7).

and maintenance of global structures in the case of some stable disks (see chapter 11).

In the main body of this monograph (chapters 1–10) emphasis will be placed on the study of essentially isolated galaxies in which internal mechanisms control all the dynamical processes of generation, maintenance, and evolution of the global spiral structures. The point of view described here is supported also by the recent infrared images of spiral galaxies, where the Population II objects stand out in regular spiral structures despite the irregular structures found in Population I objects.

0.12 Future Developments

In this monograph we shall try to show that the theory of spiral structure in galaxies has matured into a solid stage. However, it will be also made clear that many exciting issues are presently unresolved, on which interesting developments are expected, we hope in the near future. Among these issues, we should list the full theory of two-component systems (stars and gas), both in relation to the nonlinear evolution of global modes and in relation to the modeling of the processes of self-regulation and of star formation. We should also mention a full development of the theory of barred and ringed spirals of various types. Progress along these lines is expected to be accompanied by the construction of greatly improved models of individual galaxies and thus by a deeper and more detailed knowledge of galactic structure. The study of the various mechanisms involved in the development of the theory will benefit from (and possibly have a beneficial influence on) studies of collective dynamics in parallel fields of research, such as fluid dynamics, geophysics, and plasma physics. A solid grasp of the current structure and dynamics of spiral galaxies is expected to generate interesting ideas on the problem of their formation and of their long-term evolution.

0.13 References

1. Chandrasekhar, S. 1960, **Ellipsoidal Figures of Equilibrium**, The University of Chicago Press, Chicago.

2. Drazin, P. G., and Reid, W. H. 1981, **Hydrodynamic Stability**, Cambridge University Press, Cambridge.

3. Heisenberg, W. 1924, *Ann. Phys. Lpz.*, 4, **74**, 577.

4. Hubble, E. 1926, *Astrophys. J.*, **64**, 321.

5. Jeans, J. 1961, **Astronomy and Cosmogony**, Dover, New York (originally published in 1929 by Cambridge University Press), p. 332.

6. Lin, C.C. 1955, **The Theory of Hydrodynamic Stability**, Cambridge University Press, Cambridge.

7. Lindblad, B. 1963, *Stockholm Observ. Ann.*, **22**, no. 5, 3.

8. Oort, J.H. 1962, in **Interstellar Matter in Galaxies**, ed. L. Woltjer, Benjamin, New York, p. 234.

9. Orr, W.McF. 1907, *Proc. R. Irish Acad.*, A, **27**, 69.

10. Prandtl, L. 1922, *Phys. Z.*, **23**, 19.

11. Sandage A. 1961 **The Hubble Atlas of Galaxies**, Carnegie Institution, Washington, DC.

12. Zwicky, F. 1957, **Morphological Astronomy**, Springer-Verlag, Berlin.

I   Physical Concepts

Physical and Morphological Characteristics of Galaxies

Because the emphasis of this monograph is semiempirical, it is natural to start out with the description of some general observed phenomena, in particular of the various classification systems currently used for normal galaxies.

For spiral galaxies the very existence of such classification schemes argues in favor of the picture that the main spiral morphologies trace, broadly speaking, intrinsic characteristics of the galaxies and, as such, should be long-lasting and not merely the result of transient activities.

1.1 Hubble Classification

Shortly after the discovery that galaxies are large stellar systems, true building blocks of the universe located far away from the Milky Way, Hubble [11] proposed a morphological classification system for normal galaxies (see figure 1.1) that is essentially still in use today, in spite of the enormous progress made in extragalactic astronomy and of the various refinements proposed, especially by Sandage, van den Bergh, and de Vaucouleurs. The term *normal galaxies* is meant to exclude peculiar objects, which may occasionally be observed during the process of a violent interaction event and, nowadays, other extraordinary objects such as quasars or BL-Lac objects, which might even be the progenitors of normal galaxies. It is also clear that our world of galaxies is biased towards bright stellar systems because dwarf objects cannot be so easily detected and observed in their structure and morphology. Our local group of galaxies, dominated by Andromeda (M31)[1] and by the Milky Way Galaxy, has several "small" galaxies like the Magellanic Clouds

---

1. Most of the galaxies cited in this monograph are identified by their number in one of three major catalogues: (M) the Messier galaxies [13], (NGC) the New General Catalogue [8], and (UGC) the Uppsala General Catalogue of Galaxies [15].

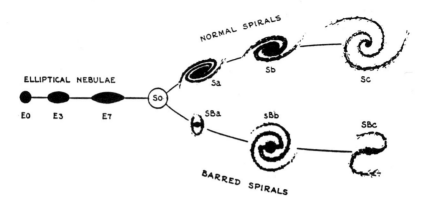

Figure 1.1
Hubble tuning fork diagram [11].

or the companions of M31; some of these are so weak sources of light
(see Leo II in the Hubble Atlas) that they are hardly recognized, and
their mass is sometimes comparable to the mass of the largest globular clusters (which are essentially spherical stellar subsystems orbiting inside galaxies—some can be seen in the pictures of M87 and of
NGC 3115 in figure 1.2).

The Gravity acts on all scales, so truly isolated galaxies do not exist. In
a first, simplified description one can distinguish field galaxies from
those that are located in small groups (such as the local group mentioned above), and from those found in large clusters (which may har-
bor more than a thousand galaxies—see the Virgo cluster or the Coma
cluster of figure 1.3). Examples of these galaxy associations are easy
to give, but in each case it is not clear whether we can fully identify
a gravitationally bound set of galaxies or, from another point of view,
whether we can decide if a given galaxy does or does not belong to the
group or cluster under consideration.

The tuning fork diagram divides normal galaxies into three main linear sequences: ellipticals (E), normal spirals (SA), and barred spirals
(SB). Ellipticals are arranged in the order of increasing flattening from
E0 (which appear round) to E7; ellipticals with flattening larger than E7
(axial ratio 3:1) are not observed. This empirical E-sequence is a convenient tool, but, unfortunately, it does not directly refer to intrinsic properties of the galaxies. An E0 galaxy, for example, need not be intrinsically spherical, because it might be a "cigar" seen from the tip or an
oblate, disklike object seen from a pole. Indeed, one of the major goals

(a)

(b)

Figure 1.2
Examples from [21] to illustrate the tuning fork diagram: (a) M87(E0) (b) NGC 3115(S0);
c-type spirals are illustrated in figure 1.12.

(c)

(d)

Figure 1.2 *(continued)*
(c) M31(Sb) (d) NGC 3351(SBb).

(e)

Figure 1.2 *(continued)*
(e) NGC 4594 (Sa/Sb).

of the study of structure and dynamics of elliptical galaxies is to recon-
struct their three-dimensional shape, based on the available photometric and kinematical data. In the absence of good data, ellipticals were
thought to be essentially devoid of gas and rotation-supported (and
thus implicitly assumed to be axisymmetric). Now they are known
to often possess significant amounts of hot (X-ray-emitting), or warm
(ionized), or even cold gas; in addition, the first spectroscopic mea-
surements of their kinematics (in the mid-1970s) revealed that their
apparent shape does not correlate with rotation, so that they should
be regarded primarily as stellar systems characterized by "anisotropic
pressure" (i.e., by different amounts of random motions in different directions) and they may often be triaxial. Progress is rapidly being made
in the understanding of the structure of elliptical galaxies (e.g., see [4]).
We shall not deal with these objects in our monograph. Here we may
just note several elements of continuity between ellipticals and spirals.
On the one hand, several ellipticals have been found to be disky, and
this may be an indication of continuity with spirals, which one might
even see in two different ways, from axisymmetric ellipticals through
S0 galaxies (see NGC 3115 in figure 1.2) to normal spirals and from triaxial ellipticals through SB0 galaxies (see NGC 2859 in figure 0.7) to

(f)

Figure 1.2 *(continued)*
(f) NGC 7743 (SBa).

barred galaxies. On the other hand, the inner parts of the spiral galaxies often have a bulge (see figure 1.4 for a picture of the bulge of our own Galaxy), which resembles a small elliptical galaxy; however, one should keep in mind that bulges appear to be rotation supported.

While the distinction between SA (sometimes denoted just by S) and SB galaxies is based on the existence of a bar in the disk, the classification along the two linear spiral sequences according to type a, b, or c is based on several criteria that refer to properties that are found to correlate along the following sequence (see table 1.1): (1) gas content (increasing from a to c); (2) size of the nuclear bulge (decreasing from

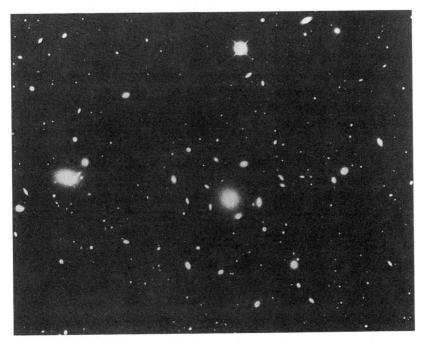

Figure 1.3
Coma cluster of galaxies [24].

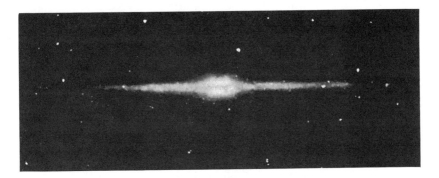

Figure 1.4
Near-infrared image of the disk and of the bulge of the Milky Way Galaxy obtained by NASA's Cosmic Background Explorer (COBE) [14].

Table 1.1
Hubble spiral sequences

$a \longrightarrow b \longrightarrow c$	
Gas content	Increasing
Size of HII regions	Increasing
Arm spacing	Increasing
Size of nuclear bulge	Decreasing
Total mass	Decreasing

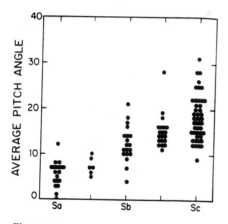

Figure 1.5
Measured pitch angle of spiral arms versus Hubble type [12].

a to c); (3) nature of spiral arms (young features, such as HII regions, are more easily recognized in late-type spirals); (4) pitch angle of spiral structure (spiral structure is tighter in a-type galaxies; see figure 1.5); and (5) total mass of the galaxy (early-types are often more massive than late spirals). (Traditionally, a-type galaxies are called "early-type" spirals, while c-type galaxies are referred to as "late-type" spirals; now it seems that there is no obvious interpretation of the Hubble classes in terms of temporal stages of one evolution process.) Generally speaking, gas content appears to be the primary parameter in the Hubble classification scheme. Note that beyond the class c, one often locates irregular spirals, which have the highest gas content. For an up-to-date discussion of physical parameters along the Hubble sequence, see [17].

We should stress that the basic Hubble classification is still applied today. More refined systems such as those adopted in recently published atlases or catalogues (RSA, **Revised Shapley-Ames Catalog of Bright Galaxies** [23]; and RC3, **Third Reference Catalogue of Bright Galaxies** [27]) tend to be more precise in locating a spiral along the linear sequences outlined above and more specific with regard to the morphology of the observed object (e.g., as to the presence of inner and/or outer rings; see table 1.2). Such refined classifications are very useful; still, we are far from producing a system suitable for an "automatic classification procedure" because of the large variety of observed morphologies in such complex systems and of the basic continuity among all the classes that can be identified as broad morphological categories. Such an overall continuity of morphologies is emphasized by the fact that the morphological type is, to some extent, dependent on the wave band of observation; that is, a bar may appear in infrared observations (which calls for an SB classification; see further remarks in section 1.3.1 and discussion and examples in chapter 5) that may have escaped unnoticed in normal optical pictures.

An important point in the Hubble classification system is that, even though the tuning fork diagram is usually drawn using, as examples, grand design, bisymmetric prototypes, no explicit reference is made in the class subdivision as to the large-scale regularity of spiral structure. We shall elaborate on this point in section 1.3.

A final remark is in order for readers not familiar with the subject. The morphological types and classes play a very important role in the following discussion because they trace some intrinsic structural properties of the galaxies. Still, the classification process is an approximate procedure, as shown by the fact that equally competent astronomers may have a divergent opinion on the classification of a given spiral galaxy, or that even the same astronomer, in the course of time, may change his or her mind on the appropriate classification of a given object.

1.2 Overall Characteristics of Spiral Galaxies

Spiral galaxies are large stellar systems, accompanied by a relatively small amount of interstellar gas ($< 5\%$ of their total mass). Their linear scale is usually in the range 1–50 kpc (1 kpc $\simeq 3.1 \times 10^{21}$cm) and their total mass in the range 10^{10}–$10^{12}M_\odot$ ($1M_\odot \approx 2 \times 10^{33}$g is the mass of the Sun). Typical velocities of stars in a galaxy are in the range 100–400 km/sec.

Table 1.2
Coding of revised morphological types [27].

Classes	Families	Varieties	Stages	T	Type	Code
Ellipticals		Compact		-6	cE	cE...
			Ellipt. (0–6)	-5	E0	.E.0.
			Intermediate	-5	E0-1	.E.0+
		"cD"		-4	E+	.E+..
Lenticulars					S0	.L
	Non-barred				SA0	.LA
	Barred				SB0	.LB
	Mixed				SAB0	.LX
		Inner ring			S(r)0	.L.R
		S-shaped			S(s)0	.L.S
		Mixed			S(rs)0	.L.T
			Early	-3	S0⁻	.L..-
			Intermediate	-2	S0º	.L..0
			Late	-1	S0⁺	.L..+
Spirals	Non-barred				SA	.SA
	Barred				SB	.SB
	Mixed				SAB	.SX
		Inner ring			S(r)	.S.R
		S-shaped			S(s)	.S.S
		Mixed			S(rs)	.S.T
			0/a	0	S0/a	.S..0
			a	1	Sa	.S..1
			ab	2	Sab	.S..2
			b	3	Sb	.S..3
			bc	4	Sbc	.S..4
			c	5	Sc	.S..5
			cd	6	Scd	.S..6
			d	7	Sd	.S..7
			dm	8	Sdm	.S..8
			m	9	Sm	.S..9
Irregulars	Non-barred				IA	.IA
	Barred				IB	.IB
	Mixed				IAB	.IX
		S-shaped			I(s)	.I.S
			Non-Magellanic	90	I0	.I.0
			Magellanic	10	Im	.I..9
		Compact		11	cI	cI

Table 1.2 *(continued)*
Coding of revised morphological types [27].

Classes	Families	Varieties	Stages	T	Type	Code
Peculiars				99	Pec	.P
Peculiarities			Peculiarity		pec	.....P
(All types)			Uncertain		:	....*
			Doubtful		?	....?
			Spindle		sp	...../
			Outer ring		(R)	R......
			Pseudo–outer R		(R$'$)	P......

1.2.1 Overall Distribution of Matter

The mass distribution appears to be roughly divided into a spheroidal nuclear bulge, which may be more or less prominent, and an extended disk. Visual or infrared images of the Milky Way show the existence of the nuclear bulge even in our own Galaxy (see figure 1.4 and color plate 1).

The mass distribution in galaxies is inferred from various plausible arguments and from the modeling of the available kinematical data (see section 1.2.3). In reality, when dealing with the overall morphology, we rely on the light distribution, which is directly observed, and on the fact that the absence of large color gradients across a given galaxy makes it reasonable to assume that the observed photometries can be converted into mass density profiles by means of a simple proportionality factor (the so-called mass-to-light ratio). (Actually, the mass of the disk is mostly contained in red stars, so that the mass distribution is best revealed in red or infrared images; see also sections 1.2.2 and 1.3.1.)

Therefore, it is important to note that the light profile of a spheroidal bulge projected along the line of sight, as for elliptical galaxies, can be approximately modeled by the $R^{1/4}$ law: $I \sim I_e \exp\{-7.67[(R/R_e)^{1/4} -1]\}$, while the stellar disk luminosity profile is approximately exponential $I \sim I_0 \exp[-r/h_*]$. Presumably, these laws also describe the mass distribution in these components. The third column of table 1.3 gives the scale length h_* for a sample of spiral galaxies for which high-quality, radially extended rotation curves (see section 1.2.3) are available. The table also records the values for the central brightness (μ_0),

Table 1.3
Parameters of galaxies with extended rotation curves [7]

Name	Distance Mpc	Scale Length kpc	D_{25} kpc	v_{max} km/sec	M_B mag	μ_0 mag/arcsec^2
DDO 154	4.0	0.50	3.06	49	-13.81	23.25
DDO 170	14.6	1.57	4.66	66	-14.95	24.60
NGC 224	0.7	5.40	31.09	252	-20.60	21.62
NGC 247	2.5	2.72	11.89	110	-18.16	22.58
NGC 300	1.9	1.80	10.78	85	-18.01	21.83
NGC 925	9.4	4.24	25.52	120	-19.84	21.87
NGC 2366	3.3	1.25	6.34	52	-16.92	22.13
NGC 2403	3.2	2.08	16.05	136	-19.26	20.90
NGC 2683	5.1	1.20	10.75	205	-18.70	20.27
NGC 2841	9.5	2.46	19.56	326	-20.31	20.21
NGC 2903	6.4	1.99	20.89	203	-19.98	20.08
NGC 3109	1.7	1.35	5.30	58	-16.84	22.38
NGC 3198	9.4	2.52	19.81	157	-19.42	21.16
NGC 3521	8.9	2.38	22.04	210	-20.49	19.97
NGC 4013	11.8	2.27	13.35	197	-19.02	21.32
NGC 4236	3.2	2.72	14.64	89	-18.24	22.50
NGC 4258	6.6	5.22	30.43	216	-20.65	21.51
NGC 4395	4.5	3.08	16.48	90	-17.83	23.18
NGC 4565	16.2	8.95	52.87	253	-21.56	21.77
NGC 4725	16.0	6.75	49.87	234	-21.38	21.33
NGC 5033	11.9	5.40	33.06	222	-20.20	22.03
NGC 5055	8.0	3.96	26.72	214	-20.59	20.97
NGC 5371	34.8	4.35	44.19	242	-21.58	20.18
NGC 5907	11.4	5.90	26.95	232	-20.20	22.22
NGC 6503	5.9	1.72	9.01	121	-18.70	21.04
NGC 7331	14.9	5.03	41.39	242	-21.36	20.72
NGC 7793	3.4	1.08	8.56	116	-18.28	20.45
UCG 2259	9.8	1.34	7.65	90	-17.07	22.13

total luminosity (M_B), and optical diameter (D_{25}), following standard astronomical notation and definitions.

The cold gas distribution is well measured by means of radio observations of atomic hydrogen (at the 21 cm hyperfine transition emission line). These radio observations show that cold gas generally extends well beyond the stellar optical disk (i.e., at radii larger than $D_{25}/2$) with a distribution characterized by a much larger scale length (than h_*). An

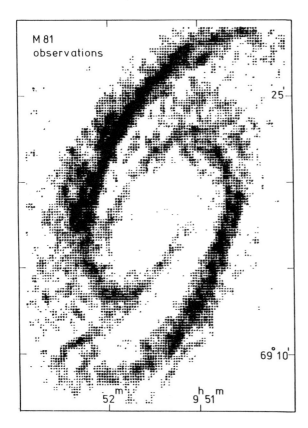

Figure 1.6
Radio image (at 21 cm) of M81 [18]. Symbols represent areas where the density of atomic hydrogen exceeds $4M_\odot/pc^2$.

example of such a gas distribution is given in figure 1.6, which shows a radio image of M81.

Another important aspect of the overall mass distribution is its three-dimensional structure. The stellar disk has a small but finite thickness, determined by the relatively low vertical velocities associated with stellar orbits. In contrast, as indicated by the study of our own Galaxy, there are "high-velocity stars" and other objects (such as globular clusters) that cannot be possibly confined to a disk, even though they do not obviously belong to the observed spheroidal bulge component. Thus it is customary to introduce the concept of a diffuse halo, which, for simplicity, is often taken to be spherical (indeed, the globular cluster system of our Galaxy has a roughly spherical distribution).

1.2.2 Stars of Various Populations, Gas, and Dust

Stars of different masses have different lifetimes. Thus a stellar disk, as currently observed, is composed of stars which formed out of the cold gas component at various epochs of time. The most brilliant stars (classified as O or B on the basis of their spectral characteristics) must have formed very recently (e.g., within the last 10 to 15 million years), so that their distribution and kinematics are similar to those of the current cold interstellar medium. Gas and young stars are therefore described as one population, often referred to as "Population I" objects. They are usually constrained in a very thin layer. In the outer parts of galaxy disks star formation appears to be turned off, and the cold interstellar medium gradually flares and/or warps because the self-gravity of the disk becomes less effective. Note that the Population I disk, being dominated by young stars of the O–B types often found in associations such as the Pleiades, is very blue.

In contrast, the older stars have a thicker mass distribution. Besides the older stars present in the bulge, or in the globular clusters, there is a fairly thick distribution of stars, which accounts for most of the mass of the disk. These older stars are referred to as "Population II" objects and are better revealed in red or infrared images. Naturally, there is a continuous variation in the age and distribution of stars, and so we should consider the separation in populations only as a convenient descriptive tool, while in reality one finds a fairly continuous transition through intermediate populations.

The interstellar medium contains several interesting components. Besides the already mentioned neutral atomic hydrogen (often denoted by HI), there is warm or hot ionized gas (especially around the so-called HII regions of recent star formation), a sizable amount of helium (in proportions determined by cosmological processes), and, in the inner disk, considerable amounts of gas in molecular form. The last are usually inferred from detections of interstellar CO, which is argued to be a good tracer of the local molecular content. The physics of the interstellar medium, as we shall indicate further, especially in chapter 3, is intimately related to the manifestation of density waves. One component that is empirically found to delineate large-scale spiral arms is dust. Dust particles (probably made of carbon aggregates) are immediately recognized to obscure the light of the stars in the Milky Way (see plate 1), so that much of the disk of our own Galaxy is opaque and inaccessible to optical studies. In external galaxies, dust shows up especially as prominent dust lanes that are found to outline the global spiral structure. But it also shows up in shorter spiral arcs interspersed in the optical spiral arms. In many galaxies the dust lanes show a very regular pattern into the center of the galaxies without being accompanied by significant numbers of young stars. Here we should also mention that dust properties have been very helpful in proving empirically that large-scale spiral arms are trailing with respect to the overall rotation of the galaxy.

Galaxies contain also magnetic fields and cosmic rays, which are likely to play a role in the dynamics of the interstellar medium. For the purpose of discussing the large-scale dynamics of the disk, it should be kept in mind that, judging from the observed conditions in the solar vicinity, the energy stored in the form of gravity and of rotation exceeds by a factor of 1,000 the other forms of energy available in these systems.

1.2.3 Quantitative Determination of Mass Distributions from Observed Rotation Curves

Galaxy disks are rotation-supported, in the sense that stellar orbits are very close to being circular. This is clearly shown by optical observations, where the Doppler shift of emission lines in the warm gas is used to derive the rotation curve, that is, the linear velocity of rotation of the disk material as a function of galactocentric radius (see figure 1.7). The study of these kinematical data shows that galaxy disks have a rather symmetric gravitational field, even when their appearance displays nonaxisymmetric features such as large-scale spiral arms. The general picture is that residual velocities of the stars, with respect to the mean rotation along circular orbits, are rather small, on the order of 10–50 km/sec. Clearly the observed velocities are giving us clues on the gravitational field of the galaxy, and thus on its mass distribution. The rotation curve $V = V(r)$ is actually better determined from radio observations of the (neutral) atomic hydrogen, that is, from the Doppler shift of the 21 cm emission line. The advantage of using HI rotation curves (see figure 1.8) is that, because of the diffuse gas distribution, they are able to trace the gravitational field usually well outside the optical stellar disk. We might expect that, outside the optical disk, the rotation curve would decline in a Keplerian manner ($V \sim r^{-1/2}$). Instead, the linear rotational velocity is found to stay fairly constant and, in some cases, even to rise slightly out to the last measured point. This is an indication that the mass distribution extends out to very large distances, much farther out than suggested by the stellar light distribution. This brings us to the problem of dark matter in spiral galaxies.

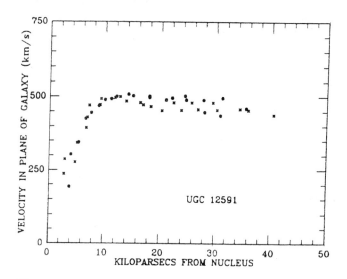

Figure 1.7
Rotation velocities for UGC 12591, the most rapidly rotating spiral disk known, from $H\alpha$ and [NII] optical emission lines; different symbols refer to opposite sides of the galaxy [19].

Before focusing on the issue of dark halos, we should stress here that it is the modeling of rotation curves in terms of the detailed support from stellar disk, gas, bulge, and halo that ultimately gives us a quantitative description of the overall mass distribution in spiral galaxies and, in particular, of the mass-to-light ratio for galaxy disks (see the examples of table 1.4). For our own Galaxy and, to some extent, also for other external galaxies, additional constraints are determined by studies of the vertical structure of the disk. In spite of the enormous progress in data collection and in the modeling of the dynamical phenomena involved, there is still a large uncertainty (by a factor of 2 or so) on the appropriate values for the mass-to-light ratio of galaxy disks (such a value may, of course, change from galaxy to galaxy).

1.2.4 Dark Matter

The problem of dark matter within galaxies dates back to the study of the vertical structure of the disk in the solar neighborhood[2] [16]. Such

---

2. The Sun is located very close to the equatorial plane of the disk of the Milky Way Galaxy, at about 8 kpc (i.e., at $R_\odot \approx 2h_*$) from the galactic center.

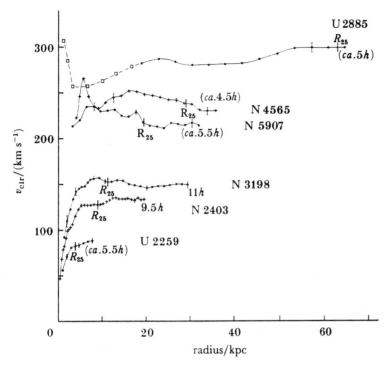

Figure 1.8
HI rotation curves for a number of spiral galaxies [2]. Distances are based on $H_0 = 75$ km sec^{-1}Mpc^{-1}. The optical radius, R_{25}, and the number of disk scale lengths, h, at the last measured point are indicated. For the inner region of UGC 2885, optical velocities have been used. All curves remain approximately flat beyond the turnover radius of the disk, $(2.5\ h)$.

investigation pointed to the fact that about half of the local mass density required in the disk by the observed vertical motions was unaccounted for by observed matter. Nowadays it seems that the vertical motions in the solar vicinity can actually be reconciled with the observed matter in the disk (see further discussion in chapter 6). Paradoxically, because this more recent interpretation of the data implies a relatively light disk, the problem of dark matter for our Galaxy is now *more* evident because in this picture at least half of the mass contained in a sphere defined by the solar circle must be in the form of a dark halo, in order to explain the observed value of the rotation velocity of the Sun around the galactic center (≈ 220 km/sec).

As described in section 1.2.3, rotation curves of external galaxies actually give the best evidence for the presence of dark halos. The discovery of flat rotation curves goes back to the early 1970s, but a proper

Table 1.4
Parameters for NGC 2403 and NGC 3198 [20]

	NGC 2403	NGC 3198	
Distance	3.25	9.2	(Mpc)
Disk scale length (h)	2.1	2.7	(kpc)
R_{25}	8.5	11.2	(kpc)
R_{max} (HI)	20	30	(kpc)
R_{max}/h	9.5	11	
V_{max}	135	157	(km/sec)
M_{HI}	3.2	4.8	$(10^9 \, M_\odot)$
M_{total}	7.9	15.4	$(10^{10} \, M_\odot)$
L_B	0.79	0.86	$(10^{10} \, L_{B\odot})$
M_{total}/L_B	10	18	$(M_\odot/L_{B\odot})$
M_{disk} (max)	1.9	4.1	$(10^{10} \, M_\odot)$
M_{halo} (R<R_{25})	1.3	1.9	$(10^{10} \, M_\odot)$
M_{disk}/L_B	≤2.4	≤4.7	$(M_\odot/L_{B\odot})$
M_{dark}/M_{lum} (R<R_{25})	≥0.8	≥0.5	
M_{dark}/M_{lum} (R<R_{max})	≥3.2	≥2.7	

focus on the problem of dark matter in spirals came much later, in the mid-1980s. Indeed, it was realized that rotation curves measured within the optical disk do not require, strictly speaking, the presence of dark halos in the sense that with an appropriate choice of a constant mass-to-light ratio it is possible to convert the observed photometric profile into a density profile compatible with the observed rotation curve. The presence of a dark halo becomes necessary when one considers the radially extended radio rotation curves. These are incompatible with the observed photometry (figure 1.9) unless one is willing to accept that the classical law of gravity no longer holds on such a large scale.

Current studies indicate that dark halos are a common feature of spiral galaxies. The shape of the halos is not known but is likely to be spheroidal. The mass in the halo is at least comparable to the amount of mass in the visible form within a sphere of the size of the stellar disk ($r \sim 4h_*$). Beyond such radius, the cumulative mass of the dark halo should grow roughly linearly with radius in order to support the observed flat rotation curve. The strength of these conclusions is based on a few detailed studies of very smooth and symmetric normal spirals. Comparable studies are not yet available for barred spirals.

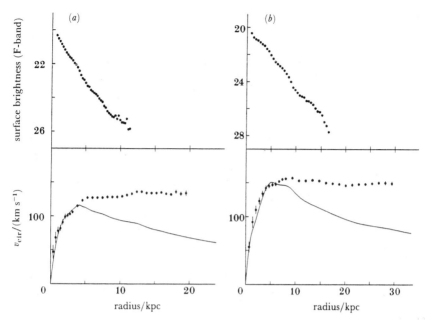

Figure 1.9
Two galaxies, (a) NGC 2403 and (b) NGC 3198, with extended, symmetrical HI disks[2]. Upper panels: luminosity profile. Lower panels: observed rotation curve (dots with error bars) and rotation curve calculated from the light profile and the distribution of HI, including a correction for helium (solid lines). The contribution of the stars to the calculated rotation curve contains the mass-to-light ratio as an arbitrary scale factor. Maximization of the disk mass (stars only), while matching the inner observed rotation curve, gives $M/L_B = 1.9$ for NGC 2403 and 4.0 for NGC 3198.

What makes dark matter? Planets? Dim stars? Black holes? Exotic particles? The problem is of considerable interest in cosmology (although the amount of dark matter "detected" by galaxy rotation curves is far smaller than the amount required in certain cosmological theories). Several observational projects are currently in progress in order to set limits on the properties of the possible constituents of dark matter.

1.2.5   Maximum Disk "Ansatz"

Once the need for a dark halo is recognized, one would like to find a procedure to decide, for the purpose of modeling a given rotation curve, how much mass has to be assigned to the dark halo and with what distribution. This procedure gives the so-called disk-halo decomposition of a rotation curve.

The problem is best exemplified by the case of Sc galaxies that do not involve significant bulges. If, in a first approximation, gas is taken to be negligible (it is actually straightforward to correct for its presence), the goal is to divide the rotation curve into two contributions,[3] $V^2 = V_D^2 + V_H^2$, where the *profile* of the contribution of the rotation curve due to the disk is fixed in shape by the observed photometry, while its *scale* is controlled by the mass-to-light ratio $(M/L)_{\text{disk}}$ assigned to the disk. The difference between the observed rotation curve and the disk contribution is attributed to the presence of the dark halo; it should be checked a posteriori that the implied mass distribution for the dark halo is not physically unreasonable (e.g., it would be desirable that the implied density profile for the dark halo decrease monotonically with radius). Obviously, the parameter $(M/L)_{\text{disk}}$ cannot exceed a maximum value, otherwise at some radius the predicted value of V_D^2 might exceed V^2. This maximum mass-to-light ratio is mostly constrained by the inner properties of the observed rotation curve (which are given by the data with some uncertainty). Thus a conservative effort to minimize the role of unseen matter leads to the choice of the maximum-disk solution for the disk-halo decomposition (see figure 1.10).

Clearly, from the dynamical point of view, solutions with a smaller value for $(M/L)_{\text{disk}}$ and a heavier dark halo are equally viable. Additional dynamical studies (for example, of the vertical structure of the disk) may actually set a *minimum* bound to the value of $(M/L)_{\text{disk}}$. There is no obvious dynamical reason why the maximum-disk solution should be the one preferred by nature. Indeed, as noted in the previous section, our own Galaxy is likely to have a massive dark halo that probably exceeds the expectations of the maximum-disk decomposition. Still, it is puzzling to find that the rotation curves, which owe their support in the inner parts mostly to the disk and in the outer parts to the dark halo, happen to be fairly flat and rather featureless in their profiles. This empirical fact argues for a (possibly primordial) tuning (a "conspiracy") of the visible and dark components. In this context, the maximum-disk solution gives a reasonable reference model of a basic state that should be clarified further.

---

3. These relations actually refer to the force balance equation for circular orbits; this is why the squares of the velocities are involved.

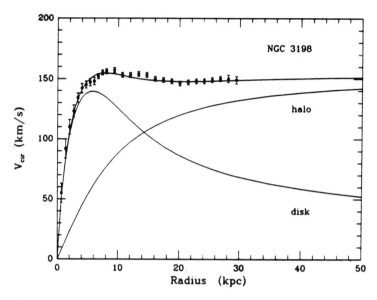

Figure 1.10
Fit of exponential disk with maximum mass and halo to the observed rotation curve for NGC 3198 (dots with error bars) [1]. The scale length of the disk has been taken equal to that of the light distribution (2.7 kpc). The maximum circular velocity of the disk has been somewhat reduced with respect to that in figure 1.9 to allow a halo with a nonhollow core.

It should be noted that the concept of maximum-disk solution can be extended to the case where other forms of visible matter (bulge, gas) are known to contribute to the observed rotation curve. In this case, the maximum-disk solution can be defined as the one that minimizes the role of unseen material.

Such a long discussion on the still-debated problem of modeling rotation curves is justified in this monograph because this is the most important step in the process of defining the properties of the basic state of spiral galaxies. In turn, dynamics are very sensitive to the properties of the assumed basic state (see chapter 7).

1.3 Morphology of Spiral Arms

Some morphological aspects in the observed spiral structure in galaxies are not addressed in the Hubble morphological classification outlined in section 1.1 and should be taken into account, and even explained, by a dynamical theory of spiral structure.

1.3.1 Morphology in Different Colors

As originally noted by Zwicky [28], the detailed morphology of spiral
arms changes significantly when a given galaxy is observed in different
wave bands. In particular, pictures taken in the blue or in the ultraviolet band tend to show very thin arms (see figure 1.11), which may
be accompanied by ragged features, spurs, branchings, and "feathers."
Very often one recognizes multiple-armed structure, which develops at
mesoscales. It should be recalled that such blue images are dominated
by the Population I component. In contrast, red and infrared images
are characterized by a much smoother and more regular spiral structure, with broad arms. Very recently, it has become possible to obtain
infrared images in the K-band (at a wavelength close to 2μ), which is
expected to be dominated by the light of evolved K and M giant stars
and thus to probe the structure of the underlying Population II disk.
Spiral structure in these K-band images is found to be smooth, regular, and generally bisymmetric; in some cases, a bar is revealed, even
when the visual picture does not show it. Thus through images in different wave bands it is possible to disentangle the different dynamical
behavior of the various populations of the disk.

1.3.2 Luminosity Classification

Especially with reference to the gas-rich, late-type spiral galaxies, van
den Bergh [3] introduced a classification system based on five luminosity classes (from I to V), in the order of decreasing intrinsic luminosity. Note that the original classification is based only on morphological
grounds (the "quality of spiral arms": class I objects have long, well-
developed, filamentary spiral arms) and that the name "luminosity
class" derives from the observed correlation with luminosity. Brighter
(and probably more massive) galaxies belonging to classes I and II display considerable regularity, with strong and sharp optical spiral arms
(see NGC 5364 and NGC 4321 in figure 1.12), while fainter objects belonging to class III or later classes have fuzzier and broader optical
arms (like M33 = NGC 598 or NGC 2403 in figure 1.12), with a smaller
degree of overall regularity. Note that all the galaxies shown in figure 1.12 are classified as Sc in the Hubble classification system. The
dispersion in morphology associated with a given luminosity class (I–
II) is illustrated for Sb galaxies in figure 1.13, taken from the Revised
Shapley-Ames Catalog of Sandage and Tammann [23].

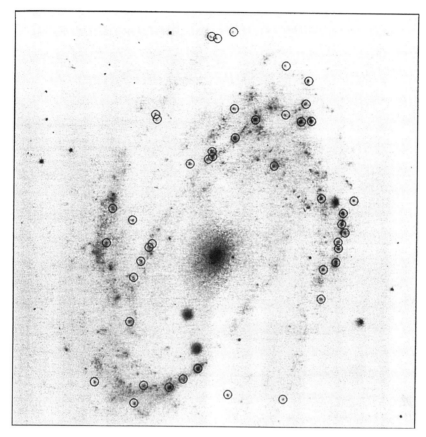

Figure 1.11
Near-UV image of M81; 46 sources identified as HII regions are circled [10].

1.3.3 Regularity Classification

In an effort to give a systematic determination of the presence or absence of a large-scale grand design in spiral galaxies, Bruce and Debbie Elmegreen [9] have introduced a regularity classification system based on twelve categories covering the range from grand design spirals like M81 to flocculent spirals like NGC 2841. Note that the large-scale regularity of spiral structure appears to be uncorrelated with the Hubble type, as indeed emphasized by the examples of M81 and NGC 2841, which are both Sb galaxies and are in many respects very similar to each other. Actually, the twelve classes of regularity appear to have been introduced with explicit reference to bisymmetric large-

(a)

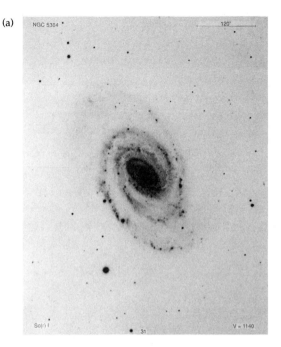

(b)

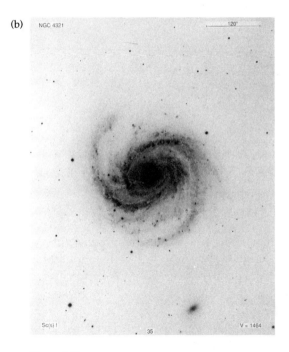

Figure 1.12
Illustration of luminosity/regularity classification: (a) NGC 5364 (Sc(r)I) [22] (b) NGC 4321 (Sc(s)I) [22] (c) NGC 5457 (Sc(s)I) [22] (d) NGC 3198 (Sc(s)I–II) [22].

(c)

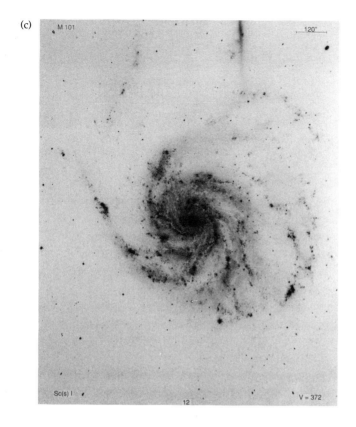

M 101

120"

Sc(s) I

12

V = 372

(d)

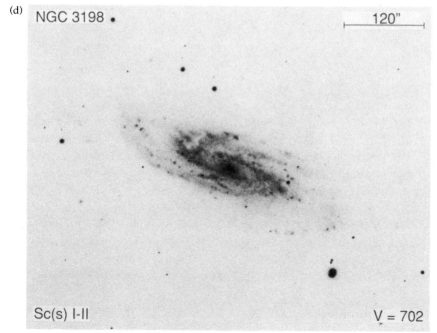

NGC 3198

120"

Sc(s) I-II

V = 702

(e)

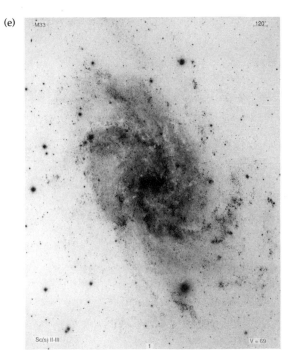

(f)

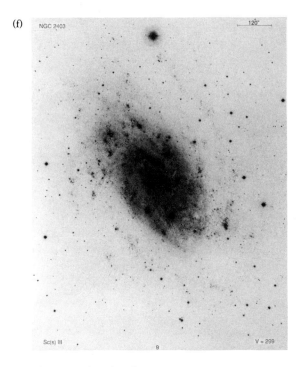

Figure 1.12 *(continued)*
(e) NGC 598 (Sc(s)II–III) [22] (f) NGC 2403 (Sc(s)III) [22].

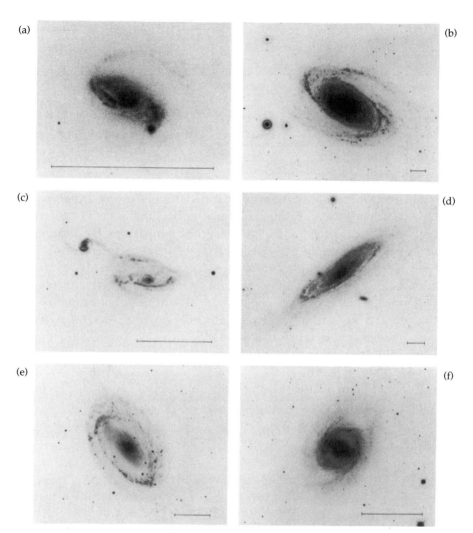

Figure 1.13
Dispersion of Sb galaxies: (a) NGC 23 (SbI–II), (b) M81 (Sb(r)I–II), (c) NGC 5395(SbII), (d) M31(SbI–II), (e) NGC 4725(Sb/SBb(r)II), (f) NGC 3351(SBb(r)II), showing that appearance does not correlate well with absolute luminosity (galaxies on the left, NGC 23, NGC 5395, and NGC 4725, are much brighter than those on the right) [23].

Table 1.5
Regularity classes

	G	M	F
Global spiral structure	Dominant	Important	Absent (or very weak)
Mesoscale spiral structure	Present	Dominant	Very weak
Small-scale spiral structure	Present	Present	Present

scale structure, while, in principle, one might conceive a very regular nonbisymmetric (e.g., three-arm) grand design structure. Attempts have been made at correlating the regularity classification with various properties, such as the presence or absence of companions, the location of the galaxy in a group, and the overall structure of the rotation curve, for a statistically significant sample of objects. Conclusive statements are not easily made or justified. Here, as an indicative characterization of the various observed morphologies, we list in table 1.5 the qualitative properties of galaxies with regular global structure (designated as grand design spirals: type G), of those without a global spiral structure (flocculent galaxies: type F), and, somewhat in between, of multiple-armed spirals (type M).

It should be reiterated that the statistical distribution of galaxies of various regularity types is bound to depend significantly on the wave band of observation, as noted in section 1.3.1. In particular, grand design spiral structure is found to be more frequent in the infrared.

1.3.4 Morphology of Bars

Bars are straight, linear features that come in a variety of morphologies. There are large bars that affect a major part of the optical disk, like in NGC 1300, or small bars that may be seen as a continuation of the large-scale spiral structure into the center, like in M83 (see figure 1.14). Very often, the existence of a bar is delineated by offset straight dust lanes. In any case, there is much continuity between barred and non-barred objects, so much that certain objects, like NGC 6951, exhibit characteristics of both SB and S systems.

Well-developed bars are often accompanied by an outer ring. The phenomenology of ring morphologies is extremely rich, as emphasized

(a)

(b)

Figure 1.14
Morphology of bars: (a) NGC 5236 = M83(SBc(s)II) [21] (b) NGC 1300(SBb) [21].

(c)

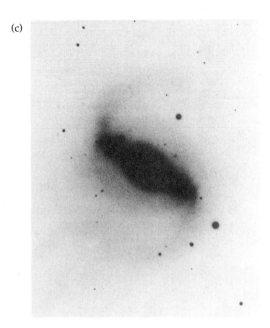

Figure 1.14 *(continued)*
(c) NGC 4314(SBa(s)Pec) [21].

recently by Buta [6]. It should be kept in mind that, even though bars tend to appear as "nonlinearly well developed" structures, in reality there is no easy direct indication of how massive the bar is. In this respect, use of images in different wave bands can give an indication on whether a given observed bar indeed has a sharp linear feature in the underlying mass distribution or, rather, is associated with a broad oval distortion of the disk.

An extreme case of barred morphology is that of SB0 galaxies (see figure 0.7). It is interesting to note that "the bar of SB0$_2$ galaxies does not extend completely across the face of the underlying lens. There are two diametrically opposite regions of enhanced luminosity on the rim of the lens which, together with the nucleus, constitute the bar"[21].

1.4   Major Issues Addressed in This Monograph

In this section we shall raise a number of issues that define the main themes addressed in this monograph. Later on, in the following chapters, we shall describe a theory, which, even if still incomplete and therefore unable to provide all the desired answers, seems to offer a promising quantitative framework for the resolution of several of the issues raised.

1.4.1 Physical Basis for the Classification of Spiral Galaxies

The main question that we would like to address is the physical basis of the empirical classification schemes for spiral galaxies. In particular, we ask the following:

• How do barred spiral galaxies differ from normal spiral galaxies? Why have certain galaxies become barred, while others have not?

• How do we explain the frequency of normal (unbarred) spiral galaxies, especially of those with a tightly wound structure?

• Why are regular grand design spiral structures generally, but not always, two-armed?

• How can we account for the observed transitions among the various morphological types and the basic continuity in these transitions?

• How can we explain the observed coexistence of morphologies within the same galaxy, especially in relation to observations in different wave bands?

• How can we explain the observed correlation among different properties within a given classification scheme (such as the Hubble diagram) and the apparent absence of correlation with other properties (such as the regularity or the luminosity class)?

• How do we explain the different degrees of regularity in the observed spiral structure? How do we explain the frequently observed flocculent structure?

1.4.2 Dynamical Modeling of Spiral Galaxies

We can identify questions, much related to those just posed, that focus more sharply on the specific aspects of quantitative modeling:

• How can we construct galaxy models that simulate categories of galaxies with a given spiral morphology?

• How can we produce a detailed model for an individual galaxy with a given spiral morphology?

- How much can we infer on the properties of the basic state of a galaxy from its observed spiral morphology? Can a proper modeling of spiral structure lead to a better determination of the intrinsic structure of disk galaxies?

Questions of this kind stress one important point. Our major goal is indeed to get to know how galaxies are structured; in a sense, this has priority over the general problems that can be formulated in the context of galactic dynamics. The determination of certain properties of the basic state can be made directly by observations combined with a simple analysis of the state of equilibrium. An example of the process is given by the study of rotation curves, as we briefly described in section 1.2. However, not only do the data have their own limitations, but even when excellent data are available (such as some accurately determined, radially extended HI rotation curves), we are often left with an intrinsic ambiguity on the underlying basic state (such as the relative share between disk and halo in the support of the measured rotation curve). Here we emphasize that a proper answer to some of the issues raised above may help establish a better determination of the basic state of spiral galaxies, given the sensitivity, when a dynamical interpretation is put forward, of the spiral morphology to the basic state that is required to support such morphology. In turn, these inferences can be taken as predictions to be tested by new or improved observations.

The modeling process is very interesting and useful but can be extremely complex. The main issue that is often very hard to prove is the dynamical self-consistency of the model produced. For example, in some models of the gas flow in barred galaxies, an oval bar mass distribution is assumed to drive the gas. Even if qualitative agreement may be obtained with some observed features, one has to show the consistency of the driving field with the assumed basic state, or in other words, one has to justify the presence of the oval bar that is assumed. A major goal of the theory that will be described in the following chapters is to propose answers to the various issues raised in relation to the morphology of spiral structure in terms of realistic self-consistent dynamical models.

1.4.3   Origin and Evolution of Spiral Structure in Galaxies

Here we come naturally to other questions that deal more directly with the problem of excitation of spiral structure:

- How are spiral structures, of the various observed morphologies, excited in galaxy disks?

- Is large-scale spiral structure in galaxies typically quasi-stationary or, rather, fast-evolving?

- If it is fast-evolving, is it a transient, occasional phenomenon or, rather, continually regenerated by internal mechanisms?

These are the questions that are usually posed first, and certainly they are most appealing to the imagination. Unfortunately, given the extremely long timescales of the phenomena involved, answers to these questions cannot be so easily checked by direct observation. Therefore, we are led to judge the viability of the possible answers especially on the basis of their impact on the morphological and modeling issues raised above (in sections 1.4.1 and 1.4.2).

Without entering a discussion that would require arguments and tools to be provided in the following chapters, we may give a broad description of some alternatives. If we refer to grand design spiral structure, broadly speaking, we may argue that its excitation is either external or internal (obviously these possibilities are not mutually exclusive, but for simplicity they are presented as alternatives).

External excitation may result from tidal interactions with another galaxy. If we take this point of view, we should be ready to address also the following questions: Why are there many isolated (field) galaxies with global spiral structures? Are appropriate encounters between galaxies frequent enough to account for all the observed global spiral structures? Are there reasonable basic states in the condition to support externally excited spiral structure, otherwise unable to generate large-scale spiral structure on their own? Intrinsic excitation would argue that tidal interactions are not required.

The prevalence of two-armed spirals has led some researchers to suggest that, even for normal spirals, the large-scale structure is in reality bar-driven; the argument would be that, for those cases where a bar is not directly observed, it is sufficient to imagine that the disk be affected by a broad oval distortion, or that a small bar be present at the center, and this would drive the observed spiral structure. If this point of view is taken, answers should be given also to the following questions: Are the central bars (like that observed in NGC 5364) massive enough to drive the main disk to form two-armed spirals? What is the mechanism for the formation of the driving bars or assumed oval distortions? Would this scenario of bar driving provide an explanation

for the gradual transition between the barred branch and the normal
branch in the Hubble classification system? A variation on the same
theme would be to invoke the action of a triaxial dark halo on the disk.
This "explanation" would appear to be ad hoc; in any case, all the relevant dynamical mechanisms should be checked to determine whether
they are astrophysically viable, and we would still have to answer all
the questions raised earlier in section 1.4.

There remains the possibility that large-scale spiral structure is internally generated and related to self-excited global modes of instability in the disk. This is essentially the line of thinking adopted in the
present monograph, a major purpose of which will be to describe the
various mechanisms that are needed and believed to be present in real
galaxies to generate and to maintain the observed spiral structures. In
relation to this point of view, we may immediately mention a few crucial issues: Would the self-excited global modes match the observed
morphologies, both in shape and in spatial scale? Wouldn't the disk
evolve, partly as a result of the invoked instabilities, with the danger
that the spiral modes would essentially quickly damp out and disappear?

This naturally brings us to the last two main issues raised at the beginning of this section. Of course the possible alternatives have to be
taken in a statistical sense because it is not inconceivable that, for some
specific objects, large-scale spiral structure is rapidly evolving, while
for others, it is long-lasting. It is also clear that in each individual case,
each scenario may be present as a secondary ingredient, for example,
in relation to structures on different scales. Naturally, each scenario
of fast-evolving, long-lasting, or regenerative spiral structure depends
on the proposed picture for the excitation mechanisms, as briefly discussed above.

The picture of a transient, fast-evolving, large-scale spiral structure is
generally associated with the hypothesis that external, tidal excitation
is responsible for grand design structure. After the encounter with an
external galaxy (as may be argued for the case of M51 being in interaction with NGC 5195), the large-scale bisymmetric structure is thought
to rapidly die away.

In a completely different picture, large-scale spiral structure is
thought to be relatively slowly evolving and long-lasting (the concept
will be explained in better detail in section 2.4 and in full depth in chapter 4), and it is taken to be the manifestation of intrinsic global modes
that are primarily self-excited and lead to the observed spiral pattern.

These modes represent the intrinsic characteristics of the basic state of the galaxy. Note that this scenario does not deny the role of interactions; for example, in the case where tidal interactions are at work, as is probably occurring for the above-mentioned system of M51 and NGC 5195, the "companion" is thought to produce modifications of the existing spiral structure, which would be likely to have some general resemblance to the one observed, possibly with less coherence. In addition, it is not excluded that interaction can temporarily give prominence to some otherwise damped modes (by analogy with the ringing of a church bell).

It is less clear how to reconcile the picture of intrinsic excitation with the scenario of a transient, fast-evolving spiral structure, as may be imagined in the regenerative point of view. Such a regenerative picture appears to be viable and interesting for *small-scale* spiral activity, but less suitable to explain large-scale coherent spiral structure.

As will become clear in the following chapters, on the basis of several empirical and dynamical arguments, we are adopting the view that large-scale spiral structure is generally quasi-stationary. We shall show how such a hypothesis can be viable and dynamically consistent and indicate how this point of view can lead to a convenient framework so as to answer most of the major issues raised in this section.

1.4.4 A Semiempirical Approach

From this fairly long discussion it should be apparent that all the issues raised are related to one another, so that we may argue that this monograph is actually pursuing one specific goal and will discuss several interesting aspects of one astrophysical problem. The adopted viewpoint is semiempirical, and thus emphasis is given to several morphological issues directly raised by the observations, while detailed dynamical mechanisms, which are hard to confront with the data, are given lower priority in the development of the theory. Very often the imagination is captured by dramatic, evolving scenarios and elegant mechanisms, which in reality are very hard to check because the data give us *current* morphologies only. In this respect *n*-body simulations are often mistaken as data, when they only reflect the dynamical properties of the adopted models.

To be sure, we do not intend to give up dynamics. We shall make good use of physical and dynamical arguments (e.g., see following discussions of the roles of gas and part III of this monograph), and we shall show that a theory can be outlined with internally consistent dynamics, but we shall try to establish confidence in it mostly on the basis of its ability to provide an answer to some key issues raised by observations (as listed in section 1.4). In this respect, we like to look at the theory in the spirit of a working hypothesis.

1.5 Relation to Other Astrophysical Themes of General Interest

Earlier in this chapter we emphasized that several issues directly raised by observations of spiral galaxies are actually different aspects of one major astrophysical problem, the problem of spiral structure in galaxies, which is the focus of this monograph. In this final section of the chapter we shall point out some of the relations between this and other astrophysical themes of general interest. Because of these relations, beneficial interactions are expected with different fields of astrophysical research.

1.5.1 Dark Matter

The modeling of the basic state of spiral galaxies requires a definition of the amount and distribution of dark matter in these objects. This touches on one of the key areas of current interest in astrophysics. Dark matter is often invoked in the cosmological context (so as to "close the universe"), but direct, convincing evidence for its presence derives mostly from studies on the much smaller galactic scale. Thus, in this respect, there is considerable interest in dynamical studies not only of spiral galaxies, but also of ellipticals, of groups and of clusters of galaxies. On the scale of clusters of galaxies, the arguments in favor of an interpretation of available data in terms of large amounts of dark matter may be less solid than those based on the analysis of the rotation curves of individual galaxies; however, they involve some of the most fascinating phenomena at the frontier of astrophysics, such as X-ray halos (as probes of the large-scale gravitational field; see figure 1.15), or the giant luminous arcs found in some clusters (which may result from images of faraway galaxies being distorted by the "gravitational lens" produced by large amounts of dark matter in the intervening cluster; see figure 1.16 and plate 2).

One obvious question is what makes dark matter, whether it is mostly baryonic (e.g., in the form of a large number of "invisible" planets) or nonbaryonic (and here a connection is directly made with

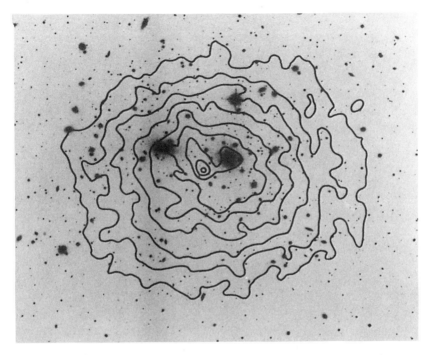

Figure 1.15
Coma cluster in X-rays; contours of constant X-ray surface brightness are shown superimposed on the optical image of the cluster [24].

particle physics and, to some extent, with what is being learned from
the experiments in large particle accelerators and colliders). Several
projects are under way to set empirical constraints on this specific
issue. Clearly the baryonic hypothesis is directly connected to the problem of star formation and to the general question of the properties of
stellar populations.

From the theoretical point of view, the somewhat elusive nature of
dark matter has spurred speculations that the laws of gravity, as known
and tested on the scale of our solar system, may even break down on
the galactic scale. If dynamics were not Newtonian, as argued by some
researchers, most of the quantitative arguments that are going to be
made in this monograph would probably fail or should be made anew.
Finally, whatever is measured or conjectured on the amount and distribution of dark matter, it has a direct impact on the various scenarios
that can be put forward regarding the problem of galaxy formation and
evolution.

(a)

Figure 1.16
Arcs in the distant cluster Abell 2218 [26] and the gravitational lens G2237+0305, the so-called Einstein Cross, where light from a distant quasar forms the four outer images because of the gravitational field of an intervening galaxy (imaged by the Faint Object Camera of the Hubble Space Telescope; credit: NASA/ESA). See plate 2.

Even without further elaboration on these points, it should be apparent that the relation with the problem of spiral structure in galaxies is only one aspect of a vast research area. Here, the study of galaxy disks seems to offer an opportunity for the determination of solid constraints.

1.5.2   Interstellar Medium and Star Formation

Another exciting area of research in astrophysics is that of the interstellar medium and of the related processes of star formation. Astronomers are working at defining a reasonable model for the interstellar medium (in our Galaxy and in external galaxies) with a proper understanding

(b)

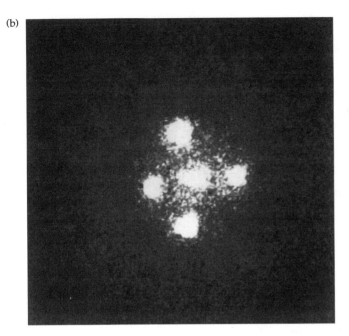

Figure 1.16 *(continued)*

of the various phases that are observed, with a description of the mass spectrum of cold clouds (from the "small" HI clouds to the "giant" molecular clouds), and with a picture of the relevant energy balance and regulation processes (including the various radiation processes in the interstellar medium and its interaction with the radiation from the stars in the formation of HII regions).

The interstellar medium is the site of the formation of young stars (see figure 1.17). One would like to know from a statistical point of view in which proportions stars with different luminosities are formed and to understand how the current star formation rates, as inferred from the data, match the observed values for the gas content in galaxies. In particular, investigations of this kind may favor the view that galaxies are currently accreting, either steadily or episodically, cold gas material from the outside. By looking at far distant galaxies, one may also get an indication of the galaxy evolution in this respect.

As will become clear in the following chapters, the properties of the interstellar medium have a major impact on the formation and on the maintenance of spiral structure, especially in normal spiral galaxies. At the same time, the behavior of the interstellar medium is very much

 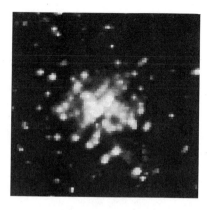

Figure 1.17
The young star cluster R136 in the 30 Doradus Nebula (in the Large Magellanic Cloud) imaged by the Wide Field/Planetary Camera of the Hubble Space Telescope (credit: NASA).

influenced by the spiral gravitational field. This indeed induces noncircular motions of the interstellar medium around the galactic center and therefore leads to collisions among the interstellar clouds and hence prompts more star formation. In the region of the minimum of the gravitational potential, the concentration of dark dust clouds and brilliant young stars is enhanced. This is the source of the well-observed dust lanes and of the brilliant young stars that appear along the arms like beads on a string. Chapter 3 will address in detail many of these issues.

1.5.3   Formation and Evolution of Galaxies

Why do galaxies possess the scales and the general properties that we observe? Why are they so distributed (in frequency) along the Hubble diagram? Theories of galaxy formation try to provide answers to questions of this kind, under a number of observational constraints among which a primary role is played by the so-called scaling laws (such as the Tully-Fisher [25] relation for spiral galaxies, see figure 1.18). These issues are of great interest because they bridge the gap between cosmological "big bang" theories and the currently observed "building blocks" of the universe. They are clearly interrelated with the problem of galaxy evolution because the elapsed time from the epoch of galaxy formation is on the order of several billion years. Then, one might even consider evolution along the Hubble sequence (for example, from Sc to

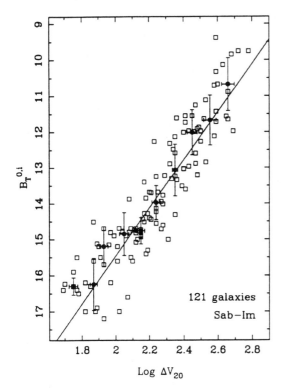

Figure 1.18
The Tully-Fisher relation [25] as defined by a sample of 121 galaxies in the Virgo cluster; for each galaxy (open squares) the absolute blue luminosity is plotted versus the width of the HI rotation velocity (corrected for inclination and extinction) [5].

Sa and S0 galaxies). When evolution is addressed, it is natural to ask whether galaxy-galaxy interaction, possibly through frequent mergers that may occur even at present epochs, plays a major role.

All of the above are tantalizing dynamical issues. Clearly, a point of contact (for spiral galaxies) with the problem of spiral structure is that the strongest constraints to the various scenarios that may be proposed come from the currently observed structure of galaxies. In turn, spiral structure is one manifestation of the properties of the basic state of these objects.

We could easily continue with other research areas of astrophysical interest, such as the search for massive black holes inside galaxies, as sometimes argued on the basis of some spectacular nuclear activity

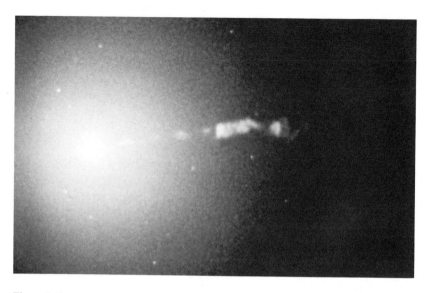

Figure 1.19
Near-infrared image of the central core and jet in M87 taken with the Wide Field/Planetary
Camera of the Hubble Space Telescope (credit: NASA).

(see figure 1.19), or the investigations of the overall chemical evolution of a galaxy. The general point we are stressing here is that a proper
theory of spiral structure should take into account a large body of astrophysical information and that, in turn, the theory of spiral structure
may have far-reaching implications even for research areas traditionally considered separate from it.

1.6 References

1. Albada, T.S. van, Bahcall, J.N., Begeman, K., and Sancisi, R. 1985, *Astrophys. J.*, **295**,
305.

2. Albada, T.S. van, and Sancisi, R. 1986, *Phil. Trans. Roy. Soc. London*, A **320**, 447.

3. Bergh, S. van den 1960, *Astrophys. J.*, **131**, 215 and 558.

4. Bertin, G., and Stiavelli, M. 1993, *Rep. Prog. Phys.*, **56**, 493.

5. Burstein, D., and Raychaudhury, S. 1989, *Astrophys. J.*, **343**, 18.

6. Buta, R. 1991, in **Dynamics of Galaxies and Their Molecular Cloud Distributions**,
IAU Symposium 146, ed. F. Combes and F. Casoli, Kluwer, Dordrecht, p. 251.

7. Casertano, S., and Gorkom, J.H. van 1991, *Astron. J.*, **101**, 1231.

8. Dreyer, J.L.E. 1888, *Mem. Roy. Astr. Soc.*, **49**, 1.

9. Elmegreen, D.M., and Elmegreen, B.G. 1982, *Mon. Not. Roy. Astr. Soc.*, **201**, 1021.

10. Hill, J.K., Bohlin, R.C., et al. 1992, *Astrophys. J. Letters*, **395**, L37.

11. Hubble, E. 1926, *Astrophys. J.*, **64**, 321.

12. Kennicutt, R.C. 1981, *Astron. J.*, **86**, 1847.

13. Messier, C. 1781, in **Connaissance des Temps pour 1784,** p. 227.

14. National Research Council (U.S.). 1991, **The Decade of Discovery in Astronomy and Astrophysics**, National Academy Press, Washington, DC.

15. Nilson, P. 1973, **Uppsala General Catalogue of Galaxies,** *Roy. Soc. Sci.*, Uppsala.

16. Oort, J.H. 1932, *Bull. Astron. Inst. Neth.*, **6**, 249.

17. Roberts, M.S., and Haynes, M.P. 1994, *Ann. Rev. Astron. Astrophys.*, **32**, 115.

18. Rots, A.H., and Shane, W.W. 1975, *Astron. Astrophys.*, **45**, 25.

19. Rubin, V. 1987, in **Dark Matter in the Universe**, IAU Symposium 117, ed. J. Kormendy and G.R. Knapp, Reidel, Dordrecht, p. 51.

20. Sancisi, R., and van Albada, T.S. 1987, in **Dark Matter in the Universe**, IAU Symposium 117, ed. J. Kormendy and G.R. Knapp, Reidel, Dordrecht, p. 67.

21. Sandage, A. 1961, **The Hubble Atlas of Galaxies**, Carnegie Institution of Washington, Washington, DC.

22. Sandage, A., and Bedke, J. 1988, **Atlas of Galaxies Useful for Measuring the Cosmological Distance Scale**, NASA SP-496, Washington, DC.

23. Sandage, A., and Tammann, G.A. 1987, **A Revised Shapley-Ames Catalog of Bright Galaxies (RSA)**, Publication 635, Carnegie Institution of Washington, Washington, DC.

24. Sarazin, C. L. 1988, **X-Ray Emissions from Clusters of Galaxies**, Cambridge University Press, Cambridge.

25. Tully, R.B., and Fisher, J.R. 1977, *Astron. Astrophys.*, **54**, 661.

26. Tyson, A. 1992, *Phys. Today*, **45**, no. 6, 24.

27. Vaucouleurs, G. de, Vaucouleurs, A. de, et al. 1991, **Third Reference Catalogue of Bright Galaxies (RC3)**, Springer-Verlag, New York.

28. Zwicky, F. 1957, **Morphological Astronomy**, Springer-Verlag, Berlin.

2

2

The Concept of Density Waves

In this monograph we shall argue that the observed spiral structure in galaxies can be best understood as the manifestation of wave patterns with a spiral gravitational field traced by the distributions of various material objects, including stars of different ages and the interstellar medium. In this chapter we shall introduce such a wave description, starting out with the following questions:

- Why do we need to resort to the concept of density waves?

- Can we explain the existence and the nature of the spiral structure on all scales with the wave concept?

- Is it natural to expect the formation of waves over the whole galactic disk?

- What are the possible mechanisms through which the waves may be generated and supported?

- Are there obvious difficulties associated with the wave concept from the observational point of view?

We begin our discussions by examining the basic kinematics and the dynamics of the stars in the galactic disk and then we continue by considering the collective behavior of the stellar system as a whole, that is, the dynamical behavior of the stars and gas clouds under the influence of the gravitational field of the total distribution of mass. In the framework developed in this monograph, a full answer to the above questions requires the introduction of the concept of large-scale global modes. Global modes, as intrinsic characteristics of the galaxy, will be described in detail later on, in chapter 4.

2.1 Density Waves versus Material Arms

Galaxy disks are rotation-supported in the sense that stars and gas, in a first approximation, move on circular orbits, rotating in such a way as to balance the gravitational self-attraction of the galactic mass that would pull them toward the center; the peculiar velocities with respect to such a general mean rotation are small. Observations show that the rotation curve $V(r)$, that is, the linear velocity of rotation as a function of galactocentric radial distance, is not proportional to the distance as in the case of the rotation of a rigid body, but rather it remains approximately constant over much of the disk (see figures 1.7, 1.8, and 1.10). Therefore, the disk is said to be in a state of "differential rotation," with the angular velocity of rotation decreasing with increasing radius ($\Omega \sim 1/r$), but not as in the case of the Keplerian motion of the planets surrounding a point mass ($\Omega \sim r^{-3/2}$, according to Kepler's law). This is naturally to be expected because the mass distribution in a galaxy is diffuse.

In such a differentially rotating system the kinematical shear would stretch any material concentration into the form of a trailing spiral arm with a pitch angle that would continue to decrease in time. Thus, if the spiral arms were material concentrations (i.e., structures always made of the same objects), over a timescale of one or two periods of revolution (typically less than half a billion years), an Sc galaxy would turn into an Sb galaxy and further into an Sa galaxy, if we were to classify the galaxy only in terms of the pitch angle of its spiral structure (see figure 2.1). On the other hand, other physical characteristics of the galaxy, such as gas content and bulge size, that are used as intrinsic parameters to judge in which Hubble category a galaxy is, cannot change so rapidly in time. This is the *winding dilemma*.

This apparent paradox can be resolved if the spiral structure is not a material arm, but part of a wave pattern. We might naively think that this explanation is contradicted by the fact that the brilliant stars and HII regions that delineate spiral arms are indeed material objects. In reality, this is not a problem because the lifetimes of these objects are very short, sometimes only on the order of one hundredth of the period of revolution. Hence, if these objects form, disappear, and reform at the crest of the waves, much like whitecaps on ocean waves (see figure 2.2a–b), the wave picture would hold. This picture does not exclude the formation of these objects elsewhere, provided that the statistical majority of them are formed in spiral regions. This point will be addressed in further detail in chapter 3.

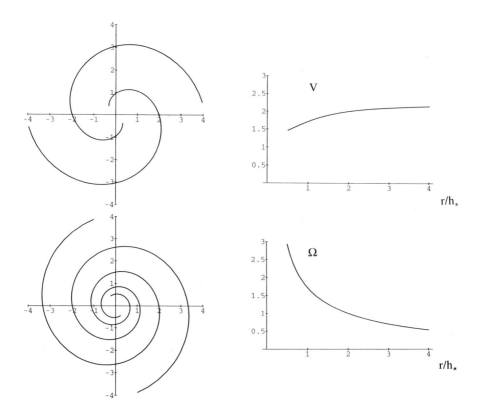

Figure 2.1
The Winding dilemma. From an initial configuration (top left, $t = 0$), material arms would be quickly stretched and wrapped up by the differential rotation (bottom left, $t = T/2$). Here T is the period of revolution evaluated at $r = 2h_*$, which for a galaxy would be typically of the order of two hundred million years. The differential rotation $\Omega = \Omega(r)$ is plotted on the bottom right frame, normalized at its value at $r = 2h_*$; on the top right frame the corresponding rotation curve $V = V(r)$ is drawn. The model used is realistic in that it includes the gravitational contribution of a bulge, of an exponential disk (of scale length h_*), and of a diffuse halo.

2.2 Tendency for the Formation of Spiral Waves

In this section we shall describe the tendency for density concentrations of stars in the galactic disk to form spiral waves. The actual process involves the mutual relation among various dynamical factors, such as the self-gravity associated with spiral arms, which will be addressed in section 2.3. However, already at the kinematical level, from the behavior of the stars in their orbital motions, it is easily recog-

(a)

(b)

Figure 2.2
Whitecaps often trace the crests of ocean waves, but they contribute less to the overall dynamics of water waves. Similarly, young and bright objects create the sharp optical features that define the arms in spiral galaxies even though their contribution to the total mass is small. [11]

nized that a galactic disk is an oscillatory system, thus prone to carry wavelike disturbances.

2.2.1   Stellar Orbits in a Galaxy

In a given axisymmetric gravitational field stars may move in circular orbits, so that the centrifugal acceleration of the individual stars is in balance with the given field. However, if all the stars moved this way, the disk would be completely cold. In reality, a physical disk will always include a certain amount of random motions, which implies deviations from circular orbits. It is easy to show that conservation of angular momentum requires that the stars perform radial oscillations about a reference circular orbit, called an "orbit of the guiding center"; such oscillations are referred to as "epicyclic motion." This statement applies if the rotation curve $V = V(r)$ is such that the function $J = rV(r)$ monotonically increases with radius.

To see this, imagine a star of specific angular momentum J_0, displaced inwards to a radius r_1 with respect to its reference radius r_0 (defined so that $r_0 V(r_0) = J_0$). Since $r_1 < r_0$, the star cannot stay at radius r_1 because its angular momentum J_0 is larger than $r_1 V(r_1)$; thus it is pulled back towards its reference radius r_0. A similar restoring force is found when one considers an outwards displacement to $r_2 > r_0$. A simple calculation shows that the stellar orbit, relative to an observer moving in the framework of the guiding center (i.e., rotating at angular speed $\Omega(r_0) = V(r_0)/r_0$), is a small ellipse described by

$$\delta r = a \sin \kappa (t - t_1)$$

$$r_0 (\delta \theta) = a \left(\frac{2\Omega}{\kappa} \right) \cos \kappa (t - t_1),$$

where a and t_1 are constants. Ω and κ correspond to the values of the functions:

$$\Omega(r) = \frac{V(r)}{r}$$

$$\kappa^2(r) = 4\Omega^2 \left(1 + \frac{r}{2\Omega} \frac{d\Omega}{dr} \right),$$

evaluated at $r = r_0$. Thus the frequency of the epicyclic oscillation is κ, which is equal to 2Ω in the case of uniform rotation.

In contrast, with reference to a fixed observer, in the inertial nonrotating frame, the orbit is a distorted circle around the center of the

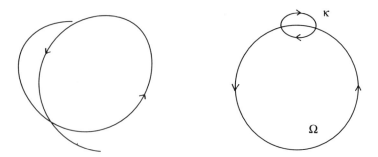

Figure 2.3
"Quasi-circular" orbit (left) in the presence of a flat rotation curve ($V(r) = $ constant; $\kappa = \sqrt{2}\Omega$); the orbit is decomposed in the motion of the guiding center with frequency Ω and of the epicycle with frequency κ (right).

galaxy, and, in general, it is not closed (see figure 2.3). Different stars found at a given location at a given time may obviously have different guiding centers.[1]

In addition to these deviations from circular motions in the galactic plane, there are also deviations in the direction perpendicular to the plane of the galactic disk. This motion is simply related to the vertical gravitational field in the vicinity of the equatorial plane of the disk. The displaced stars would tend to sink back into the disk, with a resulting vertical oscillation. For small amplitudes, the two types of oscillations are unrelated to each other. For our own Galaxy, in the vicinity of the Sun, the period of epicyclic oscillation ($2\pi/\kappa$) is shorter than the period of revolution ($2\pi/\Omega$) of the orbit around the center of the Galaxy; the vertical period of oscillation is the shortest, being shorter than the period of revolution by more than a factor of 3.

These oscillatory motions are the primary basis for the explanation of the tendency of the disk to support waves. Note that besides the observed density waves in the plane of the galactic disk there are also bending waves perpendicular to the plane of the galactic disk. It is for the theorists to demonstrate how these waves may be related in detail to the above-described oscillations in the stellar orbits.

---

1. The oscillatory motions of the stars in a galactic disk and the separation of stellar orbits into the motion of a guiding center and epicyclic oscillations around it present a number of analogies with the description of the orbits of charged particles gyrating around magnetic fields.

2.2.2   Kinematical Spiral Waves

If we focus on a specific guiding center orbit, with radius r_0, we can easily recognize that orbits in the inertial frame of reference are generally not closed because Ω and κ are not commensurate. On the other hand, we can look at the orbit in a properly chosen rotating frame (say rotating at angular velocity Ω_p) so that the orbit appears to be closed; for this purpose, we only have to make sure that $\Omega' = \Omega - \Omega_p$ and κ are commensurate. If the ratio between Ω' and κ is $\frac{1}{2}$, then the orbit closes into an ellipse centered in the center of the galaxy, much as we would expect in the inertial frame of reference if the gravitational field required a rigid body rotation (with $\kappa = 2\Omega$); the reason is, of course, that in this case during the period of two epicycles the guiding center goes around the galactic center once, so that the star goes back to its original location.

Lindblad noted that, empirically, the quantity $\Omega - \kappa/2$ can be approximately constant as a function of radius. Thus in the frame of reference rotating at such angular velocity $\Omega_p = \Omega - \kappa/2$, the quasi-circular orbits would *all* be approximately closed into ellipses! Under this fortunate circumstance, it is clear that, if at one time the stellar orbits were arranged in a set of ellipses with position angle changing as a function of radius (see figure 2.4), and if we neglect the mutual interaction among the stars along these orbits, then this arrangement of orbits could persist in time, and therefore would not be subject to the winding dilemma. In particular, the orbital configuration shown in the example of the right frame of figure 2.4 suggests a mass concentration along two spiral arms. This is indeed a wave phenomenon in the sense that stars in their motion go in and out of the regions of orbit crowding, so that the arms are seen to rotate rigidly (at angular speed Ω_p in the inertial frame of reference), which is allowed because they are made always of different stars. The coherence of the orbit crowding is only due to the special selection of orbital relations, and thus we may refer to this phenomenon as to a "kinematical wave."

From this discussion we see two clear important points of success of the picture proposed by Lindblad. In the first place, even without explaining why the arms can be formed, it opens the way to a direct resolution of the winding dilemma in terms of density waves. In the second place, the picture is viable only for structures with two arms, and indeed grand design spirals are generally two-armed (the picture would be viable for m-armed structures if $\Omega - \kappa/m$ were found to be approximately constant).

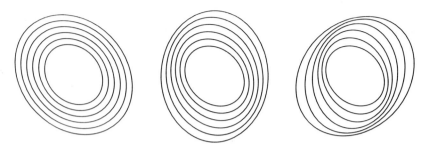

Figure 2.4
Illustration of Lindblad's kinematical waves.

On the other hand, one can list several problems with the above simple picture of kinematical waves. First of all, the quantity $\Omega - \kappa/2$ is found to be only *approximately* constant, and therefore long-term persistence could not be guaranteed. Second, following the terminology of section 2.2.1, the real disk could never be as "cold" as required in figure 2.4 in the sense that in a physical system we would have a certain amount of dispersion (random motions) with respect to the neat arrangement of orbits shown. Third, the mass concentration along the arms produced by orbit crowding would generate its own gravitational field (self-gravity), thus disturbing the orbital picture drawn here. Finally, there is no explanation of why such spiral arms would be generated, nor of their shape (which in this kinematical picture is arbitrary because the dependence of position angle of the ellipses with radius is arbitrary).

Fortunately, all of the physical ingredients missing in the kinematical waves can be incorporated in the study of the collective behavior of stars and gas, as we shall show. In particular, it will be recognized that spiral arms, when generated by such a collective behavior, cannot come in arbitrary shapes, but only with specific shapes that depend on the properties of the basic state.

2.3   Collective Behavior

On several occasions (e.g., see chapter 1) we have mentioned that in a zero-order approximation spiral galaxies can be described in terms of a time-independent basic state that is essentially axisymmetric with respect to the rotation axis. In addition, such a basic state is symmetric with respect to the equatorial plane of the disk, with $\rho(z) = \rho(-z)$.

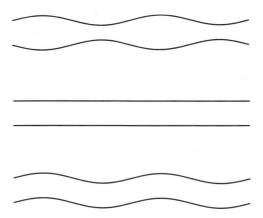

Figure 2.5
Flat layer (middle) can represent a disk seen edge-on. It may be subject to even (density; see upper part of the figure) or to odd (bending; lower part) perturbations.

It is clear that such a symmetric basic state is just an idealization of the current state, which necessarily deviates from the mathematical symmetry that is assumed. From general theoretical arguments, we could show that various types of deviation from the basic symmetry can be classified and are indeed expected to be observed. In particular, when referring to the vertical symmetry, we can consider even and odd perturbations. For the former, the density perturbation is such that $\rho_1(z) = \rho_1(-z)$, so that a density concentration in the disk is involved; for the latter, we have $\rho_1(z) = -\rho_1(-z)$, so that the disk is locally displaced vertically and a bending perturbation occurs (see figure 2.5). In either case, the perturbation involves the rearrangement of matter (stars and gas) with respect to the reference basic state; in general, the mutual forces among the parts of the system will imply a dynamical evolution of the observed deviations from the symmetry of the basic state that we can call "collective behavior." Such collective behavior has some characteristics that go well beyond the single particle orbit behavior, as described in the kinematical waves of the previous section.

In the limit of small amplitudes, even and odd disturbances evolve independently from each other. Although odd, bending perturbations have considerable astrophysical interest (see figure 2.6), in this monograph we focus our attention on the even, density-type of perturbations. For these, we shall now introduce the concept of dispersion relation, which summarizes many of the properties of their dynamical evolution (at least in the limit of small amplitudes).

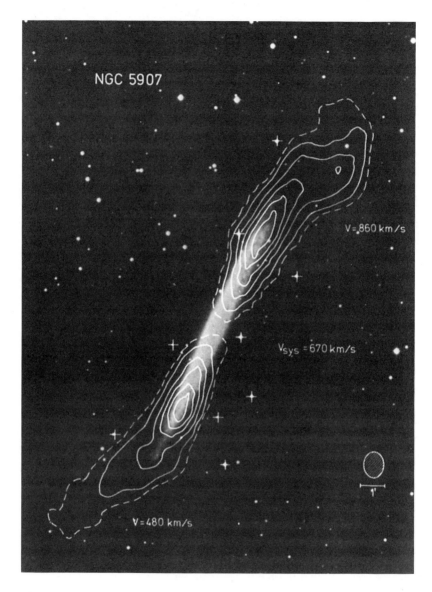

Figure 2.6
Dynamics of warped disks, like the one illustrated here in this radio map of NGC 5907
[10], can be studied in terms of odd (bending) waves.

2.3.1 Local Dispersion Relation for Density Waves

Consider a small-amplitude density perturbation. Its associated potential perturbation can in general be described as a superposition of elementary waves of the form [2]

$$\Phi_1(r, \theta, t) = \hat{\Phi}_1(r) \exp\left[i(\omega t - m\theta)\right],$$

which has a periodic dependence on the angular position θ; the integer m is usually referred to as the "number of arms," although we should keep in mind that a physical finite amplitude disturbance (called "nonlinear perturbation") generally involves many m-values even when its appearance has a given number of arms (e.g., two). Each of these elementary waves rotates rigidly around the galaxy center; indeed, if we transform to a rotating frame with "pattern frequency" $\Omega_p = \omega/m$ the disturbance appears as stationary

$$\Phi_1(r, \varphi, t) = \hat{\Phi}_1(r) \exp\left[-im\varphi\right],$$

with $\varphi = \theta - \Omega_p t$, and $\hat{\Phi}_1(r) = A(r) \exp[i\Psi(r)]$, with A and Ψ real functions of r.

Self-consistency requires that the density perturbation associated with Φ_1 (in general produced by the combined contribution of stars and gas in the disk) match the density distribution induced by the perturbed orbits as a result of the perturbed gravitational field (see figure 2.7). When combined with the appropriate boundary conditions of the problem, that is, the physical requirements at the center and at large radii that a perturbation should satisfy, the relevant equations give a global dispersion relation, that is, a relation between the allowed time behavior (ω) and the spatial structure ($\hat{\Phi}_1(r)$), of the eigenmode of oscillation, which may be growing or decaying in time. In principle, this resembles the problem of calculating the proper notes for a violin string, when the elasticity equations are studied by imposing the conditions of zero vibration at the end of the string. Such study of global modes will be discussed later on, in chapter 4.

It turns out that, in analogy with other dispersive inhomogeneous media, a local dispersion relationship can be derived for waves for which $\Phi_1(r, \theta, t)$ has a sufficiently rapid spatial variation. A general discussion of such a local dispersion relation will be given in chapter 9.

---

2. As usual, this is a complex representation—the physical quantity is described by the real part.

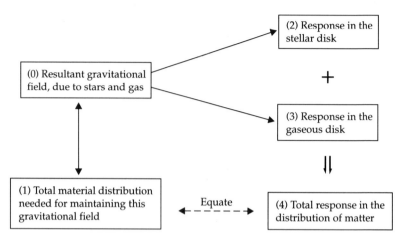

Figure 2.7
Self-consistency [6].

Here it is sufficient to refer to the relation that has been derived for the case of tightly wound spiral density waves [6]. This is the situation where the radial spacing between two successive spiral arms λ is sufficiently small so that the pitch angle of spiral structure i, given by the formula

$$\tan i = \frac{m}{kr},$$

with $k = 2\pi/\lambda$, is very small. What has just been given as a definition of tightly wound spiral waves is a description in terms of physically intuitive quantities. In reality, the proper statement (see chapter 9) refers to a requirement on the radial wave number k, defined from the relation $k \equiv d\Psi/dr$. The problem of waves with rapid spatial variations is much simpler than the general problem in that, for this case, the potential perturbation Φ_1 can be directly related, in an approximate fashion, to the local density disturbance, even though gravity is a long-range force of interaction, with the density maxima approximately in phase with the potential minima. Such a property allows for the resolution of the self-consistent problem (see figure 2.7) at each given radius. This requirement leads to the local dispersion relation, which is now a relation among ω, m, and k, through the properties of the basic state at distance r from the galactic center.

The general expression for tightly wound waves for a two-component disk of stars and gas is relatively complicated to write out explicitly [6]; it allows real solutions for the wave number and the frequency. For simplicity, we report here the much simpler dispersion relation for a one-component fluid model of a galaxy disk embedded in a spheroidal bulge-halo system:[3]

$$(\omega - m\Omega)^2 = \kappa^2 + k^2c^2 - 2\pi G\sigma |k|,$$

where Ω and κ are the angular frequency of rotation and epicyclic frequency as defined in section 2.2, c is a typical velocity dispersion (with respect to the mean rotation velocity), and σ is the disk density of mass. The quantities Ω, c, and σ depend on the galactocentric radius r and define the axisymmetric basic state. Thus, if we look at the right-hand side of the dispersion relation, we recognize the kinematical constraint required by the conservation of angular momentum (κ^2), which was described in section 2.2, a "pressure term" (k^2c^2), which would be responsible for "sound" waves in the absence of gravity and rotation, and the self-gravity contribution ($-2\pi G\sigma |k|$); the latter two terms are indeed the two key contributions missed in the analysis of kinematical waves given previously.

Note that the relation is symmetric with respect to the sign of k. It is easily realized that, if we focus on positive values of m, $k < 0$ corresponds to *trailing* spirals, while $k > 0$ describes *leading* spirals (see figure 2.8). From the observational point of view, spiral structure is generally trailing. An explanation for this fact, therefore, cannot be found in the local dispersion relation, so that one could even argue for an "antispiral theorem"; we shall show in chapter 4 that an explanation for the prevalence of trailing spirals can be found in the mechanism of excitation of global modes.

2.3.2   Spiral Patterns, Wave Branches, and Resonances

A local dispersion relation, such as the simple one for a fluid model recorded in the previous section, can be applied in two different ways.

In the first approach, it can be used in order to model an observed pattern that is argued to be characterized by a single pattern frequency Ω_p. From this point of view, the value of $\Omega_p = \omega/m$ is fixed (it is one parameter that may be determined through a best-fit analysis of certain

---

3. Here $G \simeq 6.67 \times 10^{-8}$ (cgs units) is the gravitational constant. The bulge halo is taken to be immobile and to contribute a background gravitational field, which, together with that of the disk, determines Ω and the "restoring force term" κ^2.

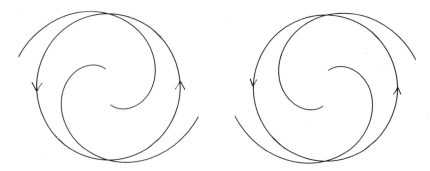

Figure 2.8
If a galaxy disk seen face-on rotates counterclockwise (as indicated by the arrows for the two cases shown in the figure), spiral arms are called "trailing" if they wind out, as illustrated on the right, or "leading" if the pattern is like the one on the left. The prevailing winding direction of spiral arms in galaxies is trailing. [13]

observational data) and, at this stage, no question is raised as to why the system selects such a value. Then the dispersion relation, in order to guarantee local self-consistency for the density wave pattern, provides the expression of k as a function of r, that is, of the allowed pitch angle $i = \arctan(m/kr)$ as a function of radius. In many galaxies the observed pitch angle is approximately constant (equiangular spiral), and the theory should be able to show why this is the case. A reference wave number is clearly given by $k_0 = \kappa^2/2\pi G\sigma$.

The local dispersion relation allows for different solutions for k at each location. For example, for the quadratic relation introduced above, we have in general two solutions for $|k|$ (i.e., at each radius we have four wave branches: short trailing, long trailing, short leading, and long leading). As is known from the theory of dispersive waves, each of these wave branches is expected to be characterized by different radial group propagation properties. These and other properties of the local dispersion relation will be described in part III of this monograph, where we shall deal with the dynamical mechanisms in detail.

We should note that, if we define as *corotation circle* the radius r_{co}, where $\Omega(r_{co}) = \Omega_p$, a good measure of the distance from the corotation circle is given by the value of the local dimensionless quantity

$$\nu = \frac{m(\Omega_p - \Omega)}{\kappa},$$

which is the frequency at which stars in their unperturbed motion would encounter the gravitational potential minima associated with

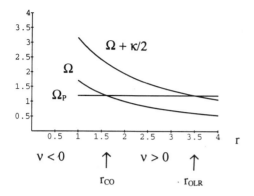

Figure 2.9
Illustration of the location of resonances in a disk in the presence of a two-armed density wave with pattern frequency Ω_p.

the spiral arms, relative to the local value of the epicyclic frequency. The wave rotates slower than the basic state inside the corotation circle ($v < 0$). When $|v|$ equals unity, the wave is in resonance with the epicyclic motion. We define as *inner Lindblad resonance* (ILR) the location, if it exists, where $v = -1$, and as *outer Lindblad resonance* (OLR) the location where $v = +1$ (see figure 2.9). The simple dispersion relation appears to be regular at the Lindblad resonances. In contrast, for a stellar disk, because of the resonance with the star epicyclic motions, one can show that waves cannot exist outside the so-called principal range of spiral structure, that is, for $|v| > 1$.

2.3.3 Jeans Stability

The second way to apply the local dispersion relation is in terms of its implied stability properties. This is best demonstrated by focusing on axisymmetric disturbances ($m = 0$). For the simple fluid model introduced above, we can ask if there is a range of wave numbers for which the disk is locally unstable ($\omega^2 < 0$) with respect to density waves, and this would occur if

$$\kappa^2 + k^2 c^2 - 2\pi G \sigma |k| < 0.$$

Then, it is useful to refer to the local marginal stability condition $\omega^2 = 0$, which here becomes

$$\kappa^2 + k^2 c^2 - 2\pi G \sigma |k| = 0.$$

It is a matter of simple analysis to show from here that, if the disk is sufficiently warm, that is, if

$$Q \equiv \frac{c\kappa}{\pi G \sigma} \geq 1,$$

then the system is locally stable with respect to all (axisymmetric) perturbations. For $Q < 1$, there is a range of wavelengths, around $\lambda \approx 2\pi^2 G\sigma/\kappa^2$, for which axisymmetric instability exists. Of course, the instability is driven by the collapse tendency characteristic of gravity and therefore is somewhat similar to the Jeans instability in three-dimensional systems. It is curious to realize that, for a disk, local Jeans instability can be completely removed (by pressure at small wavelengths and by rotation at large wavelengths).

A surprising result of the global stability analysis, which will be described in chapter 4 and in greater detail in part III, is that a disk, everywhere stable with respect to *local* axisymmetric disturbances, can be *globally unstable* with respect to global spiral modes as a result of their capability to transfer angular momentum outwards.

2.4 A Preliminary Formulation of the Hypothesis of Quasi-Stationary Spiral Structure

In 1963 Bertil Lindblad proposed the possibility of quasi-stationary spiral structure (QSSS) for galaxies with regular spiral patterns [7]. This concept was adopted by Lin and Shu [5] as the starting point for quantitative studies of the spiral structure in galaxies. They began by showing that the adoption of such a hypothesis led to a number of quantitative predictions that could be checked by observations, such as the existence of radial motions on the order of 10–20 km/sec as observed in our own Galaxy. Some of these early observational studies will be described in part II.

2.4.1 Statement

If we limit our attention to spiral structures that are *regular*, we can assume as a working hypothesis that the arms are the manifestation of a spiral pattern of density wave that, in an appropriate rotating frame, evolves only slowly.

Such a hypothesis is quite natural when applied to the large-scale structure. Indeed, we can separate local spiral structures (such as feathers extending off individual spiral arms and bridges in between two spiral arms) from global structures extending over the whole galactic disk and recognize that "although the overall impression is often hopelessly irregular and broken up, the general form of the large-scale phenomenon can be recognized in many nebulae"[9]. For example, in the two-armed global spiral structure in M81, these two kinds of spiral structures are clearly present. On the other hand, in the flocculent spiral galaxy NGC 2841, there is no indication of a global spiral structure. In multiarmed galaxies like M101, there is coexistence of global structures and local and intermediate structures.

There may be even coexistence of global spiral structures, as pointed out long ago by Zwicky [15] for M51. Recently, this point has been emphasized by studies in the infrared K-band; for a galaxy like NGC 309, the structure in the older disk component is found to be characterized by a broad bar with spiral arms departing from the tip of the bar, while the optical appearance dominated by Population I objects is more like that of a multiarmed normal spiral, such as found in M101 (see description and discussion in chapter 5).

However, we should stress that, even when a global spiral structure is present, there is no reason to believe that the structure is strictly rotating as a rigid body over many revolutions. After all, a galaxy is a complex system in which the interstellar medium is known to have sizable local structures. Binney and Tremaine [1] appear to have misinterpreted the concept of quasi-stationarity. They attributed to Lindblad and to Lin and Shu the statement "that the spiral patterns in galaxies are long-lasting, in other words, that the appearance of the pattern remains unchanged (except for overall rotation) over many orbital periods." In contrast, the hypothesis only asserts that the *global* structure for regular spiral patterns remains in a state of slow evolution; it is long-lasting only in the sense that the global structure continues to be present, and is likely to vacillate between morphologies within the same Hubble type, not that it does not change in shape. Structures on smaller scales are obviously expected to be changing much more rapidly.

Finally, we should mention that, also as a result of the torques, of the star formation, and of the heating processes associated with spiral structure, we expect the properties of the basic state to evolve on the secular timescale. Thus it should be stressed that the hypothesis of quasi-stationary spiral structure is only a tool to study the current structure of galaxies, but it does not deny, and actually fully recognizes, evolution as an important aspect of galactic dynamics.

2.4.2 Justification

The empirical basis for making the QSSS hypothesis is the existence of quite regular global spiral patterns combined with the apparent correlation between properties of the observed spiral structure and other physical properties (like bulge size and gas content) that define the Hubble sequence, so that it is unlikely a given galaxy could change its Hubble type on the short scale of the dynamical time (i.e., a couple of typical rotation periods).

The QSSS hypothesis is to be considered as a *working* hypothesis and, as such, should be evaluated on the basis of the quality of its applications (see part II). For example, some of the applications, such as the identification of certain observed features, like rings, with resonances, may turn out to provide a good empirical justification for the modeling of the arms in terms of a single pattern frequency, as suggested by the hypothesis.

Another indirect justification of the plausibility of the hypothesis derives from the experience with patterns often observed in rotating fluids (see figure 2.10).

In any case, the dynamical basis for the hypothesis lies in the existence of global spiral modes, and this in turn depends sensitively on the nature of the axisymmetric basic state, which may or may not support global spiral modes. When global spiral structures are observed, they may be expected to evolve at a much slower rate in a rotating framework appropriate to the wave pattern, provided the wave pattern is largely composed of a single dominant mode or a very small number of such modes rotating at approximately the same angular velocity. The global pattern is then, roughly speaking, a self-sustained standing wave pattern maintained by waves propagating radially in opposite directions (and rotating around the galactic center). Its general nature is very much like the wave patterns in vibrating strings, over a drumhead, or in church bells. In all these latter cases, the wave pattern eventually decays away. In the case of the galactic disk, however, it is quite likely that the wave pattern can extract energy from the huge reservoir of energy associated with the differential rotation of the system (kinetic), and with the gravitational energy (potential). Thus the wave pattern may be self-excited and eventually maintained in a quasi-steady state of evolution in the sense that the energy in the wave pattern may fluctuate slowly around a certain general level. Naturally, the wave pattern itself may evolve substantially in the course

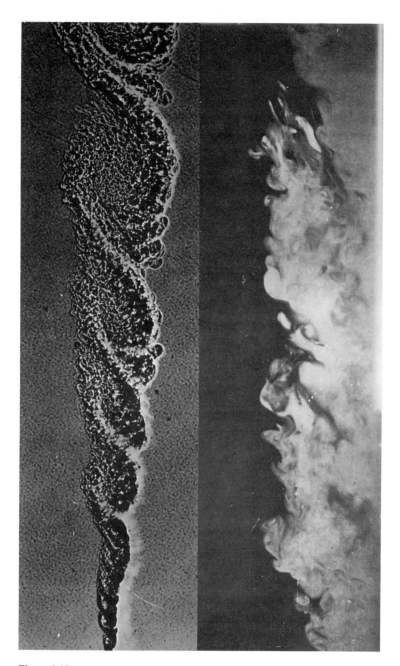

Figure 2.10
Qualitatively similar dynamical systems can display turbulent (right) or coherent (left) structures, and both extreme behaviors are predicted by the same set of dynamical equations (cf. figures 0.1 and 0.3). Indeed, in many cases both behaviors can coexist, transient and turbulent structures being usually small-scale, while large-scale patterns are long-lasting and of a grand design. The two hydrodynamic examples shown here are taken from [2].

of time, and in general it does. Thus a full discussion of the hypothesis of quasi-stationary spiral structure is postponed to chapter 4, where the concept of modes is developed and illustrated in detail. Here we should just summarize the conclusions by saying that the theory of spiral modes, in its linear form, predicts the existence of unstable exponentially growing modes as the origin of global spiral structure. Its amplitude is argued to be limited by nonlinear behavior, which is expected to become important in the interstellar medium even when the wave amplitude is fairly small (see chapter 3). It is anticipated and, to some extent, demonstrated that such an equilibration of the growing modes can take place. Again, such general equilibration of amplitude does not imply that the final global structure is strictly unchanging in time, but merely suggests the possibility of a quasi-stationary spiral structure.

2.4.3 Application

For the purpose of specific applications, we introduce two approximations: (1) the spiral pattern approximated by a uniformly rotating spiral pattern that is stationary (i.e., the pattern is at least temporarily unchanging, just as the position of a planet in the sky is temporarily unchanging at a stationary point); and (2) for simplicity of calculation, the spiral pattern approximated by a single (short) wave, satisfying the local dispersion relation. These two approximations indeed define the beginning of the density wave theory as outlined in the 1960s. Historically, its application has helped to clarify the nature of a number of astrophysical processes, such as the streaming motions of the interstellar medium, the formation of strong dust lanes, and the close association with regions of formation of young stars (see chapter 3). In particular, we may refer to the successful application to M81 carried out by Visser [14], who adopted the amplitude distribution observed in the stellar disk as an input in order to predict the gas motions; the pattern speed Ω_p was left as a free parameter, to be determined from the consideration of the general shape of the spiral pattern and of the location of the Lindblad resonances (see part II for a detailed description).

It is important not to confuse the tools for the applications with the statement of the hypothesis. If the two approximations mentioned above were taken literally, both statements would obviously be seriously flawed. It is unlikely that the pattern should be rotating as a rigid body, strictly speaking. This is the reason for the prefix "quasi":

the spiral structure is not assumed to be stationary, but only *approximately* so. Furthermore, a stationary wave pattern is in general not the result of a single wave (packet), but rather it is a standing wave pattern (much like in the problem of a vibrating string) maintained by waves propagating in opposite directions. A single (short) wave would propagate away unless continually sustained at a source. Thus, even if for practical applications the wave pattern is often approximated (successfully) by a single trailing wave of the short branch, we should keep in mind that the hypothesis of quasi-stationary spiral structure actually requires that the global wave pattern be *self-sustained*, with the help of feedback processes from the boundary conditions (i.e., the central regions and the outer disk).

2.4.4 Alternatives

In 1960 P.O. Lindblad [8] suggested, partly on the basis of his experience with numerical simulations of a stellar system, that spiral structure may be described as a "quasi-periodic" evolutionary process over a long period of time (i.e., over a period of several or many typical orbital periods). Local spiral structures are expected to be evolving approximately at the rate of shear. This is true not only for material arms but also for density waves. Powerful excitation of such evolving waves near their corotation circle was found by various authors [3,4], and this led to the scenario that spiral structures in galaxies may be regenerative (i.e., they might form, evolve, and decay to be followed by other waves). Indeed, this is expected to be true for the small-scale structures in the interstellar medium because there is expected to be a substantial dissipation of waves through turbulent motion in the gaseous clouds. Nonlinear dynamical behavior may also help to restrain the growth of the waves. For spiral structures on a global scale, the dynamical processes may be different. The "boundary" conditions near the center of the galaxy and in the outer galactic disk are important for the propagation and development of the density wave pattern over the whole disk. With a feedback process, the waves excited at the corotation zone may eventually form global modes supported primarily by the stellar component. Without a feedback process, the (short) trailing wave on a global scale is likely to propagate inwards and to wind down at the same time, resulting in a transient scenario of continually tightening spiral structure [12].

The above discussion shows the importance to distinguish carefully between a wave *pattern* and a wave *packet*, even when they may temporarily show approximately the same appearance. The former is a self-sustained standing wave with a definite angular speed, determined by the properties of the basic state; the latter propagates in both the radial and angular directions, with the angular speed as a free parameter (which happens to be generally equal to that of the pattern with the same general appearance). The amplitude distribution would be different in the two cases. For a ("standing") wave pattern, we would expect the presence of at least two waves "propagating" in opposite directions, which may interfere with consequent modulation of the amplitude distribution. In contrast, for a single wave packet, the amplitude distribution is determined by the principle of conservation of wave action. Many of these dynamical issues will be addressed in detail in part III.

2.5   References

1. Binney, J., and Tremaine, S. 1987, **Galactic Dynamics**, Princeton University Press, Princeton.

2. Dyke, M. van 1982, **An Album of Fluid Motion**, Parabolic Press, Stanford, CA.

3. Goldreich, P., and Lynden-Bell, D. 1965, *Mon. Not. Roy. Astron. Soc.*, **130**, 125.

4. Julian, W.H., and Toomre, A. 1966, *Astrophys. J.*, **146**, 810.

5. Lin, C.C., and Shu, F.H. 1964, *Astrophys. J.*, **140**, 646.

6. Lin, C.C., and Shu, F.H. 1966, *Proc. Nat. Acad. Sci.*, **55**, 229.

7. Lindblad, B. 1963, *Stockholm Observ. Ann.*, **22**, 3.

8. Lindblad, P.O. 1960, *Stockholm Observ. Ann.*, **21**, 3.

9. Oort, J.H. 1962, in **Interstellar Matter in Galaxies**, ed. L. Woltjer, Benjamin, New York, p. 234.

10. Sancisi, R. 1976, *Astron. Astrophys.*, **53**, 159.

11. Stoker, J.J. 1957, **Water Waves**, Interscience, New York.

12. Toomre, A. 1981, in **The Structure and Evolution of Normal Galaxies**, ed. S.M. Fall and D. Lynden-Bell, Cambridge University Press, Cambridge, p. 111.

13. Vaucouleurs, G. de 1958, *Astrophys. J.*, **127**, 487.

14. Visser, H.C.D. 1977, Ph.D. diss., University of Groningen.

15. Zwicky, F. 1957, **Morphological Astronomy**, Springer-Verlag, Berlin.

3   Density Waves, Interstellar Medium, and Star Formation

3.1   Processes of Star Formation

The interstellar medium is the site of star formation processes. As we have seen in chapter 1, spiral arms are generally traced by bright OB associations and young stars, which are evidence of recent and ongoing star formation. Therefore, studies of the dynamics of spiral structure have often focused on the detailed processes of star formation as a result or as a manifestation of spiral activity.

Star formation is known to take place independently of density waves and of spiral arms. Indeed, there must be intrinsic processes, most likely via cloud-cloud collisions, that account for a sizable fraction of the observed star formation rates, as is proved by the presence of young stars in irregular galaxies (see figure 1.17), or in the interarm regions of grand design spiral galaxies. It appears that the cold dissipative interstellar medium naturally generates new stars unless it is too diffuse or not settled into a thin disk (the HI distribution in the outskirts of galaxies is not associated with significant amounts of star formation). Even if the detailed mechanisms are not understood, intrinsic mechanisms of star formation are found empirically to operate essentially over the whole stellar disk.

It has also been noted [2,6] that star formation can sometimes proceed in a sort of a chain reaction that is supposed to enhance the intrinsic rate of star formation and to organize it, to some extent, into linear structures because of the presence of differential rotation in the disk. Roughly speaking, this supernova-induced star formation process consists of the following. Young stars originate a number of supernova

(a)

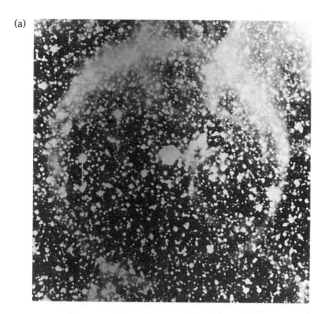

(b)

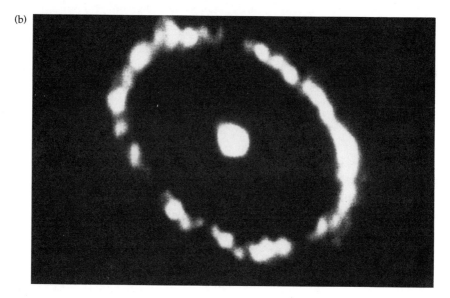

Figure 3.1
(a) Intense blast of light from Supernova 1987a, exploded in the Large Magellanic Cloud, is shown here in an image taken in 1990 at the New Technology Telescope (credit: ESO). The ring results from an "echo" on an intervening sheet of dust; its radius is approximately 1 arcminute. (b) Supernova 1987a in the Large Magellanic Cloud is shown here on a different scale, as imaged by the Faint Object Camera of the Hubble Space Telescope in 1990 (credit: NASA/ESA). The elongated ring of glowing gas, at its widest, is about 1.6 arcseconds long ($\approx$ 1.3 light years).

explosions, which can occur at the end of the short life of the most massive stars. A supernova explosion (see figure 3.1) releases a tremendous amount of energy in the interstellar medium through a shock wave that propagates in the differentially rotating cold gas. In the region swept by such a shock wave, the local star formation rate is enhanced, more young stars are created, and thus new supernovae explode, in a form of chain reaction. This mechanism is thought to give rise to streaks of bright regions in the disk (i.e., short spiral arms, with a pitch angle that is essentially determined only by the amount of differential rotation present). Starting from this basic phenomenon, it has sometimes been speculated that even large-scale spiral structure, as a sort of phase transition, can be initiated by supernova-induced star formation [3,5]. Then some observational work has tried to see whether a simple correlation can be established between the pitch angles of spiral arms and the properties of the rotation curve, in the hope of demonstrating that this is indeed a key mechanism for spiral arm formation [4]. Even if we accept that the physical intrinsic properties of the interstellar medium conform to the minimum threshold requirements for the supernova-induced star formation mechanism to operate (which is not at all clear), there is direct evidence (see especially section 1.3) that spiral structure definitely involves the underlying stellar disk, and thus it cannot be reduced just to a property of the interstellar medium. In addition, there are clear-cut kinematical signatures (see section 3.3) that provide direct simple evidence for the existence of density waves associated with spiral arms. Thus the "chain reaction" mechanism should not be regarded as the major process for the generation of large scale spiral structure, but as one mechanism of star formation especially relevant for the explanation of spurs and streaks of the kind often observed in gas-rich disks.

Finally, if the interstellar medium is swept through by a density wave (see chapter 2), the local star formation rate is expected to be enhanced as a result of the density increase. When the density wave is organized in a large-scale coherent structure as that provided by global modes (see chapter 4), then the enhanced star formation is expected to delineate the large-scale pattern (i.e., the grand design). In the 1960s such a large-scale pattern was visualized as a "shock" originating in the cold interstellar medium when the velocity of the pattern, relative to the basic state, is supersonic. This shock scenario will be sketched separately in section 3.2; in spite of its simplicity, it has served the purpose of providing a unifying framework for a number of phenomenological features. In closing this section, we would like to mention that much debate has concentrated recently on whether the density wave actually triggers the star formation processes or it just reorganizes it into a more coherent spatial pattern (see also figure 3.2). Physically speaking, one may argue that the spiral density wave enhances star formation not by adding external mechanisms, but rather by favoring the conditions for star formation already present in the interstellar medium. In turn, the overall appearance of star formation regions is also organized spatially in the form of spiral arms. From the theoretical point of view, tests of these concepts have also been attempted by means of numerical simulations of stellar disks with gas.

3.2   Sketch of the Large-Scale Shock Scenario

The following is a simple description of the large scale shock scenario elaborated in the 1960s (especially by W.W. Roberts [7]). In the pres-

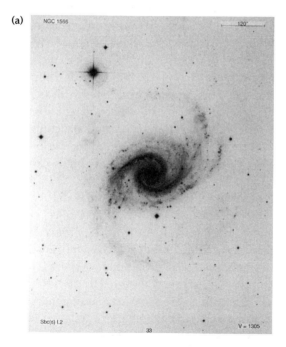

Figure 3.2
Organized and patchy star formation: (a) NGC 1566(Sbc(s)I) [11].

Figure 3.2 *(continued)*
(b) NGC 1232(Sc(rs)I) [11].

ence of a bisymmetric rigidly rotating large scale spiral field one may argue that the gas can settle into a quasi-stationary state, driven by the background gravitational field, with a response that may be nonlinear even for an imposed potential characterized by a smooth sinusoidal azimuthal structure, if the relative motion between the spiral wave and the cold interstellar medium is supersonic (i.e., far from the corotation circle). The gas thus settles into distorted oval-shaped flux tubes with a kink in the vicinity of the imposed potential minima, where a shocklike structure develops.

The gas response can be approximately calculated in terms of a "local" solution defined by the properties of the basic state and of the imposed spiral field at a mean location r from the center of the galaxy (defined as the average radius of the flux tube). Therefore, the main parameters that control the behavior of the gas properties along the flux tube are the pitch angle, the angular speed (in particular the relative local speed of the pattern with respect to the medium, i.e., how far the flux tube is from the corotation circle) and the strength of the bisymmetric spiral, and the local value of the turbulent "sound" speed of the cold gas component. In a first approximation, where the dissipation related to the resulting shock is considered to be small, it has been shown that stationary shock solutions for the gas response equations (mass and momentum conservation equations) can indeed be found, and these display astrophysically interesting properties (see figures 3.3 and 3.4). If the spiral field is strong enough and sweeps the gas medium rapidly enough, along the flux tube the gas is seen to enter the region of the potential minima with a supersonic velocity component orthogonal to the spiral pattern and to leave such a region with a subsonic orthogonal velocity component. Just before the location of the potential minima the gas suffers a sharp compression, by a factor of 5–10 in density.

On the basis of this solution, one can argue that the observed spatial structure of spiral arms should be arranged in the following way. A dust lane should mark the location of the sudden HI compression. Beyond this dust lane (i.e., moving towards the outside of the spiral arm for a trailing spiral structure inside the corotation circle) one should find the spread-out HI arm. Allowing for a delay of up to a few 10^7 years between the compression of the gas and the resulting star formation process, one should find HII regions and OB associations and other relatively young stars displaced from the dust lane according to their age. Indeed, the large-scale shock scenario was much inspired by observations of the relative location of the various components of the spiral arms and served as a useful predictive tool to explain the observed morphology of spiral structure, especially in relation with the interstellar medium. Great success has derived from the application of these concepts to interpret the fact that dust lanes are generally observed on the inside part of the spiral arms, which is natural if the corotation circle is located in the outer parts of the optical disk. In some cases (especially for barred galaxies, but also for a few normal spirals like UGC 2885 [9]), the properties of dust lanes and HII regions have been taken as evidence that the corotation circle of the observed structure may sometimes fall more inside, within the optical disk.

The examples calculated in the 1960s and the cases often shown in the literature usually refer to models and to parameters taken to represent a galaxy like the Milky Way. We should be aware, in this respect, that these phenomena depend both on microphysical processes (such as dissipation, cloud-cloud collision, star formation, ionization, and stellar evolution) and on dynamical processes (the organization of the large-scale shock and the dynamics of spiral structure). In general,

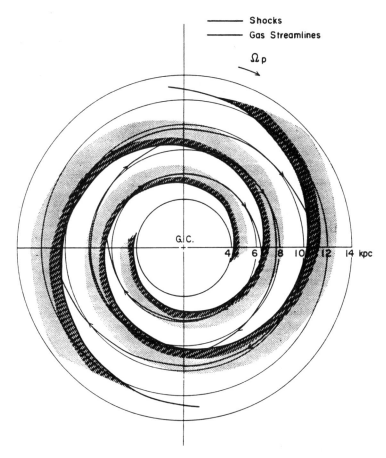

Figure 3.3
Expected structure in the presence of a large-scale, two-armed, trailing shock inside the corotation circle (gas streamlines are arrowed) [7]. A sharp HI gas peak and a narrow dust lane should mark the location of the shock on the inner side of the bright optical arms (as delineated by the newly formed young stars and HII regions).

there are mechanisms that occur on the very short timescale and others that involve the longer dynamical timescale. For this reason, the "stationary shock" calculation may be a good representation of the resulting physical structure of spiral arms, even when the underlying spiral field is not so symmetric and slowly evolving. This is likely to apply in even stronger form to relatively large objects (like UGC 2885) for which the relevant dynamical timescale is very long, and to be less imposing on the conditions of smaller spirals (like UGC 2259), for which the

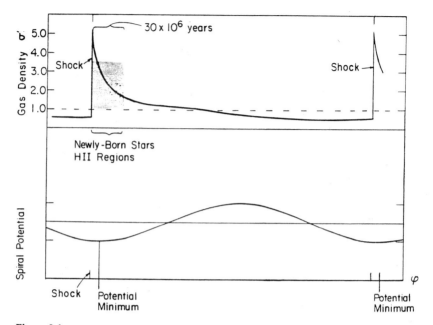

Figure 3.4
Diagram illustrating smooth field, large gas density contrast, and sequence of features encountered along a typical streamline of figure 3.3, in a (distorted) circle, while crossing spiral arms [7].

overall spiral appearance may be less regular just because the relevant timescales are not so well separated from each other.

An immediate concern, raised when the large-scale shock scenario was proposed, was that the stationary shock would be untenable because dissipation at the shocks would make the gas rapidly collapse to the center of the galaxy. Again, it is a matter of timescales. There is no doubt that dissipation, much like in accretion disks, is bound to generate an inflow of matter towards the center. However, there is little or no empirical evidence for dramatic effects in this respect. In addition, from the theoretical point of view, dissipation in the shocks has been found to provide a welcome saturation mechanism for the large-scale spiral modes so that the otherwise exponentially growing spiral modes can equilibrate at finite amplitudes.

Another source of concern, often raised as an issue, is the ability of the gas to undergo a nonlinear response and the adequacy of the description in terms of a spiral field imposed (by the stars) over the passive gas component. It had been immediately recognized in the 1960s that one should consider the resultant spiral field, produced by the combined action of stars *and* gas (see figure 2.7). Thus a proper description, which is so far *not* available, should include the full nonlinear, two-component study where gas and stars play an active share in the support of the spiral pattern. Therefore, at least in its quantitative aspects, the large-scale shock scenario as computed in the 1960s should be regarded just as a useful first approximation of the much more complex actual physical situation. On the other hand, empirically there is little doubt that the interstellar medium is undergoing some kind of large-scale shock phenomenon (see also section 3.3).

Finally, an interesting discrepancy has been pointed out recently between this scenario and the observed sequence of kinematical structures in spiral arms, for galaxies such as M83 that have been studied in great detail ([1]; see also chapter 5). These discrepancies, which occur in systems that exhibit large-scale coherent behavior, appear to indicate that something may be incorrect in our basic concepts regarding the interstellar medium. In other words, we may be forced to revise our modeling of the cold interstellar medium in order to find consistency between the dynamical theory of spiral structure and some observed properties of spiral arms.

Calculations of the gas circulation and shock formation have been performed also in the context of "bar driving" (see figure 3.5). One of the main successes of these studies has been to provide a plausible explanation for the origin of the offset dust lanes often observed in the central regions of barred galaxies.

3.3 Direct Simple Evidence of Density Waves

The interstellar medium gives the best direct evidence for the existence of density waves. It was immediately realized that, if indeed spiral arms are density concentrations that follow the dispersion relation described in chapter 2, then mass conservation would require a well-defined behavior of the motion of the gas in the vicinity of the observed spiral arms. In other words, the velocity field in a spiral galaxy should also display a wave pattern, with a well-defined phase relation between the "kinematical spiral arms" and the density spiral arms. In addition, because the gas, being cold, has a sharper, nonlinear behavior, as we saw in the section 3.2, the "shock" structure should show up as a nonlinear corrugation in the overall velocity field. The size of these

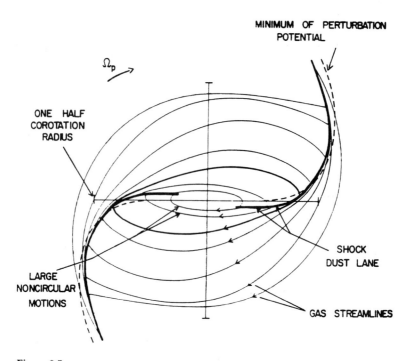

Figure 3.5
Gas circulation and shock formation in the presence of a field that includes an oval, barlike central distortion [8].

velocity perturbations over the otherwise smooth axisymmetrical flow is directly determined by the amplitude of the density wave (i.e., by the strength of the underlying spiral field). Actual density amplitudes are not easily determined observationally because one has to convert the observed *light* enhancement into a *density* enhancement, and the mass-to-light ratio is one of the parameters that we would like to determine. A similar comment applies to the gas density distribution, which is usually detected in the form of HI (atomic hydrogen), while the total number of molecules present is not so easily determined. Therefore, in this respect, the kinematical spiral arms give an even better determination of the amplitude of the wave present because they are fairly easily measured and directly give the strength of the spiral density wave.

A first indication of the presence of density waves in our galaxy was indeed given by the observed kinematics around the main arms and the observed wiggles in the overall rotation curve. Much of this work, which historically has played a major role in the development of the theory, will be reviewed in chapter 6. In chapter 5 we will also give a fairly detailed discussion of other relevant observations in external galaxies. Here, with the help of a few figures, we would like to give a preview of these observational studies, which, even without any mathematical discussion, offer direct simple evidence for the presence of density waves.

Radio studies of external galaxies in HI are one major source of information. In figure 3.6 we show the wiggles in the "radial" velocity (relative motion of the atomic hydrogen in the line-of-sight direction) for M81. The corrugations, found along the location of the optical spiral arms, are immediately apparent and conform to the picture outlined in section 3.2.

Rotation curves of external galaxies have been determined by studying the optical emission lines produced in HII regions of ionized gas. According to the scenario of section 3.2, HII regions are found prevalently along one of the sides of the dust lanes of spiral arms. The resulting effect on the optical data of the rotation curve is exactly what is observed (see figure 3.7).

External galaxies have been recently studied in CO, which is generally assumed to be a good tracer of the molecular hydrogen distribution and should also outline the location of star formation regions. In color plate 3 we reproduce one set of data for M51, which also gives a direct indication of the large-scale organization in the form of a density wave.

Finally, the large-scale shock scenario predicts that if star formation is enhanced at the potential minima associated with spiral arms, one would be able to identify age-groups of stars that are progressively older as one moves away from the wave front, because of the drifting associated with stellar orbits. Sandage indeed points out that, even in a non–grand design galaxy like M33, such age-groups can be identified, so that the stars themselves give a direct simple evidence of the passing wave front.

Certainly, the interstellar medium is an excellent indicator of the underlying structure and dynamics of spiral galaxies. It has been mostly through observation of the interstellar medium that the density wave theory has been tested and had its major impact in offering a framework for studies of galactic structure, although for several years (especially in the 1970s) the gas had almost been relegated to a role of tracer of the dynamics of spirals, thought to be mostly dominated by the stars

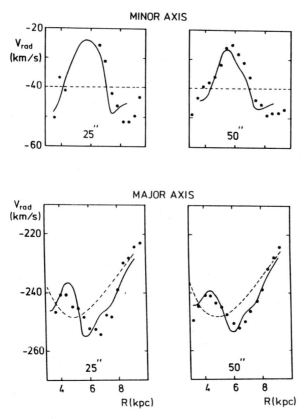

Figure 3.6
Wiggles in HI [12]. Gas velocities along the line of sight for the minor and major axes of
M81 at 25″ and 50″ resolution. Dashed lines represent the velocities of the axisymmetric model. Dots are observed data points, while full lines refer to a model that will be
described in chapter 5.

(partly because the gas component accounts for only a small fraction
of the total mass of the galaxy). In section 3.4 we shall show, instead,
that the gas plays a crucial *active* role, especially for normal spirals. Recognizing its influence naturally resolves some of the apparent puzzles
encountered in the development of the theory.

3.4   Mechanisms of Self-Regulation

We have mostly touched on only two of the dynamical roles played by
the gas. We have already remarked that the gas component is a source
of excitation for spiral structure because it is very cold. Random mo-

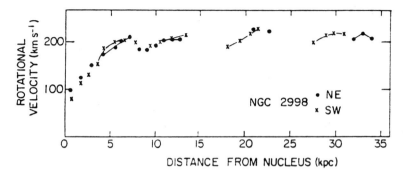

Figure 3.7
Wiggles in HII [10]. The optically measured rotation curve for NGC 2998 shows positive velocity gradients in regions of strong emission across spiral arms. Note the fairly good velocity agreement between velocities from NE and SW major axes.

tions in the cold gas component are in the range of 4–8 km/sec without sizable variations from galaxy to galaxy, while the local density ratio $\alpha = \sigma_g/\sigma_*$ of gaseous to stellar matter can be easily larger than 15% in the outer disk ($r > 2h_*$). Thus, because of the relative abundance of gas, the outer disk is "dynamically cool." Note that the local Jeans instability favored by gas can be "saturated," so that the outer disk may be marginally Jeans-stable at the *local* level. However, this marginal state allows for the transfer of waves (across corotation) toward the outermost regions so that *global* modes (see chapter 4) can be excited. The second dynamical role of the gas has been briefly mentioned in section 3.2, namely, its role as a natural mechanism (via the large-scale "shock" dissipation) to equilibrate the linearly unstable growing modes at finite amplitudes.

These two roles have been recognized and studied in the last three decades, although modes and global structures were often investigated in the context of purely stellar disks (not only analytically, but especially through *n*-body simulations), as if the gas played a secondary role and only modified a physical scenario essentially dictated by the stellar disk alone. From these stellar dynamical studies a "perennial heating" paradox gradually emerged as a major worry (if we envisage that large-scale spiral structure is generally associated with self-excited spiral modes). Stellar disks are subject to perennial heating (as shown even in exaggerated form by *n*-body simulations); thus spiral structure, even if present initially, would rapidly die out. Note that such perennial transformation of ordered kinetic energy into random motions for the stellar disk is expected to be produced not only by internal mechanisms (such as scattering of star orbits by giant molecular clouds) but also by external tidal interactions; these are indeed likely to contribute to the heating of the stellar disk, which would act later against the formation of stellar spiral arms. From this point of view, the conditions for the internal excitation of spiral structure appeared to be fairly precarious.

At this stage it was realized that the dissipative stellar medium has a third highly beneficial dynamical role, that is, its role as a kind of "dynamical thermostat" for the disk. A posteriori, this sounds very natural, if we recall that observed large-scale normal spiral structure, except for a few rare cases of smooth-armed spirals, is *always* associated with gas and Population I objects. The important point is that the Population I thin disk is not only there to trace the spiral field but plays a major active role in supporting the conditions for the existence of such spiral structure.

The dynamical thermostat is a process of self-regulation that occurs as a result of the coupling between the stellar component and the gas component of the disk. Therefore, in order for this self-regulation to occur, the stellar disk must be relatively cool at the beginning (although by itself it can be quite stable with respect to the onset of Jeans instability). In this respect, the process is not expected to be relevant for barred galaxies (which owe their bar instability to the excess in mass of the disk and should be characterized by a fairly warm stellar disk—see section 4.5) nor for flocculent galaxies (where there is little or no evidence for the presence of organized stellar spiral structure—see section 4.6).

If the velocity dispersion of the stars is not too high, the local condition for Jeans instability is shared by the two components, the stars (which carry more of the mass but are less active because of their relatively large velocity dispersion) and the gas (which carries less of the mass but is colder and more active). Now we may assume that the stars are heating up on "secular" timescales of a few revolution periods, as is consistent with many observational facts and theoretical arguments. On a very short timescale, however, the gas component is *dissipative*, so that it tends to cool off by itself unless it is stirred by instabilities. Therefore, the modest heating in the stellar component can be easily compensated by a very small decrease of the turbulent velocity of the gas in such a way that the combined disk is kept marginally stable with respect to local Jeans instability. The combined system cannot become too stable with respect to Jeans instability (otherwise, the gas would rapidly cool and make it less stable), and it cannot become unstable (otherwise rapid stirring due to Jeans instability itself would heat up gas and stars to bring the system into a more stable condition). As was mentioned before, this mechanism acts as a thermostat on the *local* stability conditions. But these are just the conditions that are required for the system to be (moderately) *unstable* at the *global* level with respect to large-scale spiral modes (see section 4.2). (The properties of this mechanism have also been demonstrated by the study of a simple set of nonlinear model equations for the time evolution of a two-component disk.)

On the timescale of several periods of revolution, the increased random motions in the stellar disk will become too high and the stellar disk will become effectively decoupled from the gas and mostly inactive. Spiral activity will then rely only on the properties of the gas (which will be subject to its own internal mechanism of regulation), so that only small-scale spiral structure will survive. According to this scenario, all spirals are bound eventually to become flocculent spirals (see section 4.6).

This self-regulation scenario should be tested and modified with respect to two important effects related to the "environment" (i.e., tidal interaction and especially gas infall from the outside). It has sometimes been argued that spiral galaxies are currently accreting sizable amounts of gas from the outside, although available X-ray data do not encourage this picture. On the other hand, gas may be accreted on a more occasional basis, in big lumps at a time, as has been argued from observing galaxies like M101. Certainly, in either form, fresh gas can "revive" spiral activity in a galaxy; this may be a welcome way to replenish the cold layer that is continually consumed to form new stars.

In concluding this chapter, we should stress that the important effects associated with the presence and with the distribution of the cold interstellar gas are to be incorporated properly when the relevant modeling procedure is developed, in order to identify physically realistic basic states for dynamical analysis (see chapter 7).

3.5 References

1. Allen, R.J., Atherton, P.D., and Tilanus, R.P.J. 1986, *Nature*, **319**, 296.

2. Blaauw, A. 1964, *Ann. Rev. Astron. Astrophys.*, **2**, 213.

3. Gerola, H., and Seiden, P.E. 1978, *Astrophys. J.*, **223**, 129.

4. Kennicutt, R.C. 1981, *Astron. J.*, **86**, 1847.

5. Mueller, M.W., and Arnett, W.D. 1976, *Astrophys. J.*, **210**, 670.

6. Opik, E. 1953, *Irish Astron. J.*, **2**, 219.

7. Roberts, W.W. 1969, *Astrophys. J.*, **158**, 123.

8. Roberts, W.W., Huntley, J.M., and Albada, G.D. van 1979, *Astrophys. J.*, **233**, 67.

9. Roelfsema, P.R., and Allen, R.J. 1985, *Astron. Astrophys.*, **146**, 213.

10. Rubin, V.C., Ford, W.K., and Thonnard, N. 1978, *Astrophys. J. Letters*, **225**, L107.

11. Sandage, A., and Bedke, J. 1988, **Atlas of Galaxies Useful for Measuring the Cosmo-logical Distance Scale**, NASA SP-496, Washington, DC.

12. Visser, H.C.D. 1977, Ph.D. diss., University of Groningen.

13. Vogel, S.N., Kulkarni, S.R., and Scoville, N.Z. 1988, *Nature*, **334**, 402.

4

Regularity, Morphological Classification, and the Concept of Spiral Modes

4.1 Global Modes as Intrinsic Characteristics of the Galaxy

In spite of the great variety of morphologies and scales that are observed, the existence of well-established classification schemes such as the Hubble diagram supports the view that some form of "law and order" governs the current state of spiral galaxies. Therefore, galaxies are thought to have reached a state of quasi-equilibrium, of which the observed spiral morphologies should trace the intrinsic characteristics (in particular, the structure of the overall gravitational field and the distribution of density and random motions for the stars and for the gas). In this respect, it is reasonable to assume that the properties of the large-scale spiral structure, which define one of the main criteria at the basis of the Hubble classification scheme, are not subject to rapid changes over the dynamical timescale (see section 7.2.6).

By analogy with other collective dynamical systems that are studied in hydrodynamics or plasma physics, one may start out by identifying a basic state as a zero-order approximation of the system under investigation and by interpreting the large-scale deviations from such a basic state as the manifestation of global modes. In this framework the basic state is chosen to have the highest degree of symmetry and smoothness compatible with the real system and is usually time-independent.

For the case of spiral galaxies, the basic state can be described as an axisymmetric disk of stars and gas embedded in a spheroidal bulge halo (see chapter 7). This procedure turns out to be very useful. From the theoretical point of view, one can study how global modes originate from a given basic state and thus provide an "explanation" for the observed morphologies. From the observational point of view, one would like to separate out the "smooth background" (i.e., the basic state) from the data for modeling purposes and so that proper input to and comparison with the theory can be made. Note that, in their effort to model spiral galaxies, astronomers routinely make use of the concept of basic state when they refer to quantities such as the rotation curve (which is a measure of the smooth gravitational field) or the HI (atomic hydrogen) density as a function of radius; in fact, all these quantities are defined and obtained by fitting a zero-order basic state to the data. From both the theoretical and the practical viewpoint this separation becomes nontrivial when the perturbation on the basic state is highly nonlinear (i.e., when the real system is too far from axisymmetry). An example of this situation is given by galaxies that possess a strong, well-developed bar. For the description of these cases a first analysis in terms of perturbations over an axisymmetric basic state may still be qualitatively valuable, but quantitative progress would require defining as a basic state a reference system that is nonaxisymmetric from the beginning. For example, in the study of many barred galaxies one should try to look directly for a basic state where the disk contains an oval distortion. Unfortunately, the construction of self-consistent, nonaxisymmetric equilibrium solutions is a very difficult and so far unresolved subject.

A given basic state is generally subject to perturbations in the form of waves and instabilities, the dynamics of which are regulated by the various forces acting on an element of the system when slightly displaced from its basic equilibrium state. Waves and instabilities typically propagate as signals across the galaxy disk; because of this, on the small spatial and temporal scale, the disk is expected to display rapid variability. If, however, we look at the disk over larger and larger scales, the elementary "signals" involved will have had room and time to see the boundaries of the system and the inhomogeneities of the disk, with the possibility of setting up coherent, essentially "standing wave" patterns. In particular, signals are bound to be refracted back ("feedback") from the central regions, especially when a bulge is present. On the other hand, signals propagating outward are likely to be absorbed as a result of a resonance (Lindblad resonance) with the local epicyclic frequency, or directly into the interstellar gas. Thus large-scale "modal" behavior is expected to take over on the large scale, while transient behavior should dominate only on the small scale. Therefore, starting from generic initial conditions, the large-scale spiral structure, as a departure from the axisymmetric equilibrium basic state, should be described in terms of a few global modes, much like the oscillatory state of a church bell or a violin string can be represented in terms of a few intrinsic notes. Like a standing wave on a violin string, each global spiral mode can be visualized in terms of the superposition of elementary density waves propagating radially in opposite directions.

This interpretation of the observed large-scale spiral structure is of direct astrophysical interest. Because many structural properties of the galaxies cannot be directly measured or can be determined only approximately from observation, spiral structure as a manifestation of global modes provides a "dynamical window" (i.e., it can be used as a dynamical probe to set constraints on properties of the basic state such as the mass-to-light ratio for the stellar disk). It is on this basis that one hopes to obtain a dynamical interpretation of the empirically founded Hubble classification of spiral galaxies (see section 4.7).

A proper definition of the concept of global modes and of the dynamical mechanisms involved requires careful mathematical analysis. This will be sketched in part III, especially in chapter 10; chapter 11 will address some of the subtle issues that can be raised in the dynamical context. In this chapter, we shall instead briefly describe the physical aspects of the modal theory and its astrophysical impact.

4.2 Excitation Mechanisms

Global spiral modes can be self-excited, that is, they can be generated naturally in galaxy disks without any external driving. Indeed, there is plenty of energy stored in the form of differential rotation that can be released and justify the excitation of spiral modes.

The basic amplification mechanism is called "overreflection" and takes place when the disk is moderately stable, so that wave signals can propagate across the disk; it occurs in the *corotation zone*, that is, in the annular region of the disk where the wave signal approximately corotates with the basic state around the center of the galaxy. Typically, inside the disk defined by the corotation circle the disturbance rotates more slowly than the underlying disk, so that it is associated with a negative angular momentum density. The corotation zone acts like a refractive boundary in such a way that when a signal reaches it from the inside it generates a transmitted wave that propagates out to the outer regions of the disk and a reflected signal that returns back towards the galaxy center. If a net amount of angular momentum is transported outward, in the process the reflected signal is amplified (i.e., overreflection occurs) in the sense that it has a larger amplitude with respect to the initial signal that entered the corotation zone. This readily shows that, if a feedback from the central regions is available, a global instability can take place, with a net gain per cycle reminiscent of the laser process. In practice, one can identify two basic forms of this "waser process," the first one occurring when a long trailing wave is overreflected into a short trailing wave signal [8], the second one, often referred to as "swing amplification" [15], which occurs when a leading signal is converted into a higher amplitude trailing wave. The source of excitation for the mode can thus be traced to its coupling a negative energy region with the outside positive energy density region of the disk (see figure 4.1). The quantitative aspects of these mechanisms can be calculated in detail and the results can be compared with the direct numerical integration of the basic equations.

Overreflection processes have been studied in the context of meteorology in recent years; they also bear some resemblance to the processes that generate vorticity islands, or "cat's eyes," in shear flows (see figure 4.2), or magnetic islands in sheared magnetic configurations (see figure 4.3). The general basic principle is that energy stored in the form of shear can, under the proper circumstances, be released by collective modes.

The faucet through which the energy is tapped is controlled by the physical parameters that define the corotation zone. A key factor that controls the energy release for the excitation of spiral modes is the amount of random motions. If the disk is too hot in the corotation zone, it is inactive, the coupling of the inner disk with the outer regions is prevented, and incoming wave signals cannot be properly amplified.

From the preceding discussion it becomes immediately clear why gas is so important, dynamically speaking, for the excitation of spiral structure. First, the gas, being a cold component, favors the overreflection process because it tends to reduce the random motions effectively present at the corotation zone. Second, the gas is also cooling, so that it can maintain these conditions longer even when competing heating processes are present (see section 3.4). Finally, in the very outer regions of the galaxy the abundant gas favors the absorption of outgoing signals, thus cooperating in the net outward transport of angular momentum across the galaxy disk. It is for these reasons that the corotation circle in unbarred spiral galaxies is expected to be located in the outer parts, where the gas becomes important relative to the stellar compo-

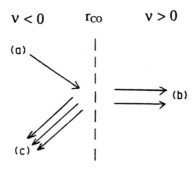

OVER-REFLECTION AT COROTATION

Type A	(a)	long trailing wave
	(b)	short trailing wave
	(c)	short trailing wave

Type B	(a)	open leading wave
	(b)	open trailing wave
	(c)	open trailing wave

Figure 4.1
Two types of overreflection. In regime A of lighter disks, the mechanism amplifies an incoming (relative to the corotation zone) long trailing wave into a stronger short trailing wave ((c), returning back toward the galaxy center); it is at the basis of the excitation of normal spiral modes. In regime B of heavier disks, a leading wave is amplified into a stronger trailing wave; this is at the basis of the excitation of barred spiral modes.

nent (figure 4.4), while, in barred galaxies, the role of gas is judged to be of secondary importance (see section 4.5).

When we move away from the corotation zone, the angular speed of the spiral wave pattern, relative to the local angular speed of the basic state, can be large enough to produce a condition of resonance with the local epicyclic frequency. The locations where these conditions are met are called "Lindblad resonances" (see section 2.3.2). In the Lindblad resonance zones wave energy is absorbed by the basic state. Wave

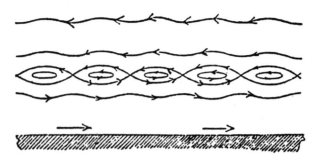

Figure 4.2
Formation of "cat's eyes" in a shear flow [7].

energy is completely absorbed in the case of a purely stellar disk and
only partially absorbed in the case of a fluid disk. Thus Lindblad resonances act like sinks that play an important role on wave signal propagation and on wave cycles for global modes. In particular, the outer
Lindblad resonance, which occurs at a location outside the corotation
circle, participates in enforcing the outer boundary condition that we
have described above (sometimes referred to as "radiation boundary
condition"). On the other hand, whenever an inner Lindblad resonance
occurs for a given wave signal, the related absorption cuts off the possible feedback, so that it damps the mode that would have otherwise
been created. This effect is found to be particularly efficient against
multiple-armed modes or modes characterized by a very low pattern
speed (see chapter 10). In conclusion, whereas the outer Lindblad resonance favors the excitation of spiral modes, the occurrence of an inner
Lindblad resonance *limits* the number of global modes that can be excited on a given disk.

4.3   Linearly Growing Modes, Moderately Growing Modes, Nonlinear Equilibration

A linear global stability analysis of realistic models of galaxy disks
shows the possibility of basic states with a discrete spectrum of moderately unstable modes. The linear analysis, that is, the study of very
small perturbations, has much in common with the study of small oscillations for a simple mechanical system, such as a pendulum. It leads
to important information on the ways the system evolves away from its
basic state, but this information is obtained by a description of the evo-

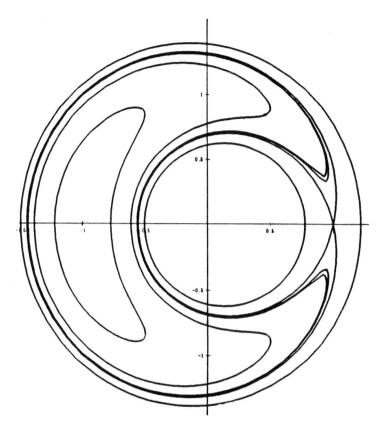

Figure 4.3
Formation of magnetic islands in a sheared magnetic configuration as a result of processes of magnetic reconnection [6].

lution of the real system that is only approximately correct. In particular, the unstable modes that are found only give a trend indicating in which direction the basic state is likely to evolve (i.e., the way in which axisymmetry is expected to be broken). Such trends are described in quantitative form, so that one can match observed spiral structure with the spatial structure of the modes. Still, one should not be misled into thinking that the real system is actually evolving at an exponential rate, because nonlinear effects are expected to take over quickly and to equilibrate the system into a new, nonaxisymmetric state. For what will follow (see especially section 4.4), it is important that only a few modes with moderate growth rate are dominant.

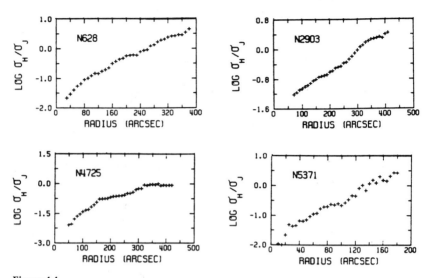

Figure 4.4
Local gas-to-star density ratio is measured in terms of observed fluxes in the radio (atomic hydrogen) and in the optical (J emulsion, stellar light), showing that the gas usually has a fairly diffuse, radially extended distribution [16]. The data shown refer to two Sb galaxies (lower frames) and to two Sc galaxies (upper frames). Pictures of NGC 628 and NGC 4725 are given in figures 0.2 and 1.13. A picture of NGC 2903 is available in the Hubble Atlas [11].

It should be noted that, on the basis of the symmetry properties of the relevant equations, it can be shown that linear global modes that would be purely oscillatory in time (i.e., neither amplified nor damped) are bound to have a "barlike" spatial structure (i.e., no well-defined spiral winding). In contrast, growing modes do have a natural winding direction and this is found to be trailing, as "required" by observations. Thus the interpretation of observed spiral structures in terms of nonlinearly saturated unstable global modes is particularly appealing.

The spectrum is discrete in the sense that only a few global structures, with well-separated pattern frequencies, dominate the long-term dynamical evolution of the system. This is so because the boundary conditions, much like the ends of a vibrating string, allow for only a few selected "notes." In addition, the Lindblad resonances severely limit the total number of linearly growing modes (see section 4.2).

In the picture that we are advocating, we should make sure that the modes involved are only *moderately* growing. Indeed, the presence of fast, violently unstable modes would clearly tell us that we have made the wrong choice of basic state; in such a case, the system under investigation would be subject to rapid evolution on the dynamical timescale and thus would be unsuitable to describe the *current* basic state (of spiral galaxies) that we are trying to model.

From the physical point of view, the interaction of the modes with the cold component of the interstellar medium is expected to provide a natural equilibration mechanism (see previous descriptions in sections 3.2 and 3.4). When the amplitude of spiral structure is sufficiently high, dissipation in the gas (which may take place also in the form of star formation events) can balance the growth generated by overreflection, and the modes are said to "saturate at finite amplitudes." In a dynamical system, one may think of several other distinct mechanisms participating in such a nonlinear equilibration process, although it may not be easy to work out the various analytical aspects of the processes involved. One possibility is that at higher amplitudes stellar orbits fail to provide the necessary support to the self-consistent gravitational potential. Interaction among the linear modes can also be a cause of nonlinear equilibration. It may just be that at a certain stage (e.g., for barred galaxies) a new quasi-equilibrium state is naturally reached. These are areas of research where useful progress can be made in the near future.

In an effort to study under controlled conditions the nonlinear evolution of basic states taken to represent galaxy disks, several types of *n*-body simulations have been performed. Indeed, in many cases evidence has been found for "modal behavior." On the other hand, the results of *n*-body simulations can easily be misinterpreted. For a correct interpretation of the results of *n*-body simulations of spiral structure, great attention should be paid to the choice of the basic state in the initial conditions; to the excessive noise and relaxation often present; to the limitations of the code to simulate detailed effects in phase space, such as those related to the Lindblad resonances; to a proper inclusion of the gas component (especially in view of the resulting nonlinear equilibration process); and to the overall reliability of the code to properly describe the dynamics of the disk over a relatively long timescale. Actually, for technical reasons, *n*-body simulations often adopt a gravity law that is different from Newton's law (for example, by introducing a so-called softened gravity). All these factors remind us that we are still far from having at our disposal reliable "experiments" for testing global theories of the dynamics and evolution of galaxy disks.

4.4   The Hypothesis of Quasi-Stationary Spiral Structure

At this stage we can proceed to complete the statement of the hypothesis of quasi-stationary spiral structure (see section 2.4). The hypothesis refers to the underlying structure, on the large scale, which, for regular and for grand design spiral galaxies, is expected to be dominated by one or two global spiral modes (equilibrated at finite amplitudes). The resulting *appearance* may thus change on the small scale fairly rapidly, and on the large scale may vacillate between morphologies of the same Hubble type even on the dynamical timescale, although the underlying *structure* is thought to be quasi-stationary.

Therefore, as originally argued when the theory of spiral structure was initiated, in order to have a first quantitative, although approximate, description of the physical processes involved, one can proceed by fitting a single wave to the observed patterns. Such a fit can be performed easily, well before a comprehensive theory of spiral modes is developed; on the other hand, when such a fit is studied, one already has in mind that the pattern is actually *self-sustained*, and that the use of a single wave is just a convenient approximation to the actual dynamical situation.

In this spirit, the hypothesis of quasi-stationary spiral structure naturally applies to the older disk (Population II) and only partially to the younger Population I thin disk. For this latter component, even when large-scale structure is present and quasi-stationary, much activity is expected on the small scale, where the quasi-stationary concept fails or, at least, is not useful. What we are really arguing is that the large-scale *gravitational potential*, in regular and grand design spirals, is quasi-stationary.

In order to give dynamical support to the above hypothesis and to establish a dynamical window on the structure of the spiral galaxies, which is a much more useful step than the single wave fit to spiral patterns, it is necessary to demonstrate the existence of families of realistic basic states in which the global stability analysis shows the dominance of only a few moderately unstable global spiral modes. Basic states subject to too many modes or to violently unstable modes would not be characterized by a quasi-stationary spiral structure. Such a viability of the modal approach has indeed been checked in detail and is at the basis of the results that will be illustrated in sections 4.5, 4.6, and 4.7 [2,3].

Before moving on to the illustration of the modal interpretation of spiral morphologies in galaxies, we should stress once more (see section 2.4.1) that the hypothesis of quasi-stationary spiral structure has often been misinterpreted. On the one hand, the prefix "quasi" has often been overlooked as if the statement claimed perfect stationarity or steadiness, which is clearly physically implausible, or as if it implied changes only on the "secular timescale" of several billion years, which may apply to the properties of the basic state but is not meant to hold for the observed morphology. On the other hand, the hypothesis has often been taken to exclude time variations (and spatial irregularities) on the small scale, while it has been clearly stated from the beginning that transient (small-scale) and quasi-stationary (large-scale) structures can well coexist within the same galaxy.

4.5 Morphology of Modes: Normal Spirals versus Barred Spirals

In order to assess the viability of the modal approach, we carried out a large survey of the properties of basic states to be used as models of disk galaxies and of their intrinsic global modes [2]. Our survey proceeded along four basic steps.

In the first step we followed the properties of a representative two-armed global spiral mode as a few key parameters were changed that control the profiles of the three functions characterizing a fluid model of a galaxy disk (i.e., the rotation curve, the disk mass density, and the equivalent acoustic speed). These profiles were chosen within the general form suggested by observations (see chapter 7). The global stability analysis involved was carried out numerically for hundreds of different galaxy models.

In the second step we retained only the models for which the representative mode was only moderately unstable, so that we discarded all the models that would be rapidly evolving because of intrinsic violent global instabilities. Therefore, we were left with a set of basic states subject to linearly unstable spiral modes that are expected to be soon equilibrated at finite amplitudes.

The third step consisted of a thorough study of the physical basis of the models thus selected. The fluid model adopted in the survey is just a useful representation of a much more complex physical situation, where one should recognize the role played by the three-dimensional distribution of matter and by the presence and distribution of cold interstellar gas. For example, when the cold gas is actively involved in the dynamics of the disk, the scale length of the density distribution of the basic state is larger than the scale length of the stellar distribution because in the outer parts the gas becomes more and more important (see also figure 4.4). Similarly, the role of gas should be properly taken into account when the profile for the equivalent acoustic speed of the fluid model is chosen (see also section 3.4). This third stage serves as a crucial step in order to establish a proper correspondence between the models of the survey and the actual physical objects that are going to be "simulated."

In the final step we made sure that the "representative mode" used in the survey was indeed the expected dominant mode for the various selected models and that the basic states under study did not possess too many modes. Basic states that are subject to too many modes could be a model for less regular or even flocculent spiral galaxies, but certainly could not be used to interpret grand design spiral galaxies in this approach.

The main results of this large survey of models and modes are summarized in figure 4.5, which illustrates the properties of four prototypes of models and of the related global spiral modes. Starting from the lower left, and then moving clockwise, we show a model characterized by a normal spiral mode. The disk that supports such a mode is relatively light (i.e., the ratio of active disk to total mass within the optical radius is less than $\frac{1}{3}$) and the corotation circle for the mode falls at 3–4 exponential scale lengths of the stellar disk. The models on the top left (figure 4.5a) and the top right (figure 4.5b) represent heavier and warmer disks that are bar-dominated. The two-blob structure does indeed lead to a bar when the basic state (subtracted out and not shown in frames a and b) is filled in (figure 4.5e). For these barred systems, the corotation circle falls slightly outside the tip of the bar at 1–2 scale lengths of the exponential stellar disk. The final model, on the lower right frame (figure 4.5d), in spite of its "nice" spiral appearance, should be generally discarded because the mode shown has a high growth rate. For this latter case, the supporting basic state is at the same time too massive and too cold; such a system would rapidly evolve into a different (warmer), more stable basic state, possibly with a bar.

The transition between stable and unstable models with respect to bar modes is in many respects reminiscent of the well-known phenomenon of the symmetry break for the classical ellipsoidal figures of equilibrium (see figure 4.6), as already recognized by Ostriker and Peebles [9]. When large amounts of angular momentum are carried by the

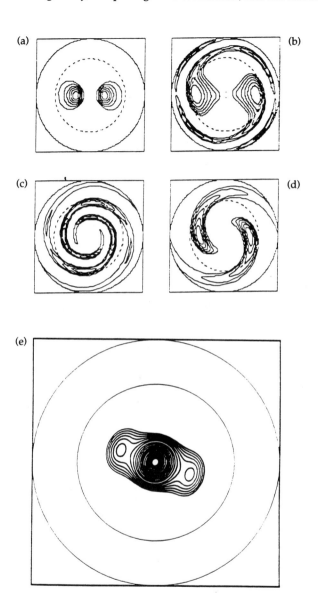

Figure 4.5
Mode prototypes [2]. Four key morphological types are compared: (a) SB0, (b) SB(s), and (c) S, all with moderate growth; and (d) a violently unstable S mode (lower right corner). The contours represent positive perturbed density in arbitrary units, in steps of 1/7 of the peak value, from the 1/7-contour upward. Dotted circles identify corotation. The frame at the bottom (e) shows the superposition of a mode of the SB0 type onto its basic axisymmetric mass model.

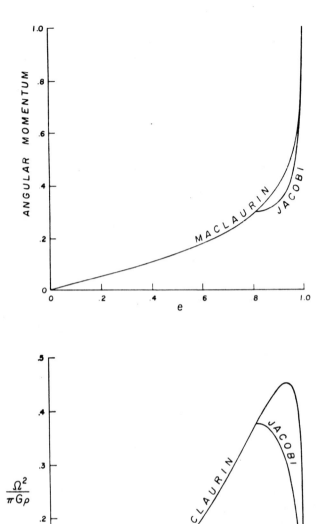

Figure 4.6
Ellipsoidal figures of equilibrium for self-gravitating, incompressible, rigidly rotating fluids [5]. At high values of the angular momentum, the triaxial (Jacobi) sequence bifurcates from the axisymmetric (Maclaurin) sequence. The quantity $e = \sqrt{1 - a_3^2/a_1^2}$ is the eccentricity of the ellipsoid. Maclaurin ellipsoids that are too flat ($e > 0.81267$) are unstable with respect to bar modes.

disk (as is the case when the disk is heavy), a bar configuration ("tri-axial ellipsoid") may be energetically favored and axisymmetry would tend to be naturally broken, in a kind of "phase transition" (see [4]).

4.6 Grand Design versus Flocculent Galaxies

One of the outstanding issues related to spiral morphology concerns the observed range of regularity in the large-scale spiral structure, from flocculent to grand design morphologies (see figures 1.12 and 3.2). Empirically, it appears that the underlying stellar disk participates in the structure, especially in grand design spirals and to some extent also in multiple-armed spirals. Objects that are gas-rich may appear patchy and less regular in the blue, and smoother and coherently organized in the red, where the bisymmetric structure becomes more prominent.

From the dynamical point of view, this different behavior can be easily interpreted. Whenever cold interstellar gas is available, small-scale and not so regular spiral structure is expected. Such a small-scale "chaotic" spiral activity may be the only possible structure if the stellar disk is too warm, so that stellar orbits cannot get organized into a global spiral mode. In this case, where the stellar disk is too stable and unresponsive, we have flocculent spiral structure. In contrast, if the stellar component is cool enough, it can be responsive and partic-ipate in the multiple-armed spiral structure (which is still mostly gas-driven), or, for galaxies like M81 that are less rich in gas, the stars may play a dominant role on the large-scale spiral structure in the support of a well-organized global two-armed spiral mode. Thus grand design spirals are interpreted as those cases where gas and stars are well cou-pled dynamically to each other (see also section 3.4). This discussion refers to the case where the disk is not too heavy; for heavier disks, a bar is expected and a bisymmetric spiral is "driven" in the gas.

We note that this picture easily explains why bisymmetry is so com-mon among grand design spirals, which indeed are star-dominated disks, even if gas plays an active role through self-regulation. In the stellar disk, multiple-armed modes are suppressed by inner Lindblad resonance (see section 4.2) and only a few two-armed and possibly one-armed modes are expected to be excited, thus supporting a genuine quasi-stationary spiral structure. If the gas content is too high, even in the presence of a well-organized, two-armed, star-dominated mode, multiple arms are easily generated.

Some of this behavior is neatly reproduced in figure 4.7, which illustrates a set of simulations, from grand design spirals that are

(a)

(b)

Figure 4.7
Simulations of spiral activity in the gas in the presence of an underlying two-armed global spiral mode of various strengths [10]. Frame (d) is a simulation run in the absence of the underlying mode.

(c)

(d)

Figure 4.7 *(continued)*

realized when a global mode dominates the scene, to less regular gas-dominated spiral structure. Figure 4.8 focuses on a set of well-known late-type spirals that display a morphological behavior that parallels the simulations of figure 4.7.

Galactic encounters may in some cases reinforce the bisymmetry and the coherence of spiral structure (as in M51—see figures 0.4 and 0.6) in objects that, because of their gas richness, would otherwise display a multiple-armed spiral structure (like M101—see figure 1.12).

4.7   A Tentative Interpretation of the Hubble Sequence

On the basis of the modal description of galaxy disks, we can thus propose the following dynamical framework for the morphological classification of spiral galaxies. Three main physical factors are recognized (see figure 4.9): the gas content (relative to the stellar content), the thin disk mass (relative to the thicker or spheroidal mass), and the "temperature" of the stellar disk (relative to that required for marginal axisymmetric stability). On the axes suggested in figure 4.9 one should imagine typical values that assess the relative importance of these factors on the overall dynamics, rather than any specific number that may be thought to be measured at a given radius. This qualification is required because we are dealing with highly inhomogeneous and complex systems that cannot be reduced to a three-dimensional parameter space.

The transition between normal and barred spiral morphology is thought to be related mostly to the disk mass factor. At high disk mass (see figure 4.10) the stellar disk should be fairly warm and a bar mode, possibly evolved into a strongly developed nonlinear bar, is expected to dominate the scene. The observed morphology would change from SB0 for gas-poor disks, to later types of barred spiral structure, up to cases like NGC 6951, where gas is so plentiful that it starts to compete with the morphology controlled by the stellar disk. At these high values of disk mass the stellar disk cannot remain cool even if we think it may have originated that way. Cool heavy stellar disks are violently unstable and observed disks are unlikely to be in such a state. Empirically, the only cases where we may argue that a relatively cool and massive stellar disk is present are a few examples of smooth-armed spiral galaxies (see NGC 7743 in figure 1.2), but they are not expected to possess a long-lasting spiral structure.

(a) (b)

(c) (d)

(e) (f)

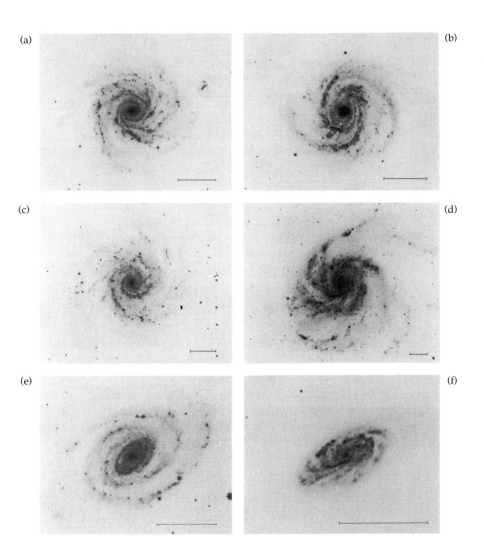

Figure 4.8
Set of Sc galaxies illustrating different degrees of patchiness and regularity of spiral arms
[13]. Pictures of NGC 1232, 4321, 628, 5457, and 5364 have been shown earlier in the
monograph. The lower right galaxy (f) is NGC 3294.

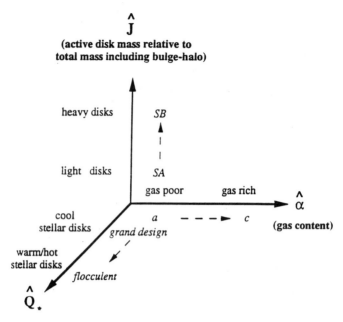

Figure 4.9
Framework for the classification of the morphologies of spiral galaxies on the basis of their intrinsic modal characteristics (see [1]).

At lower values of disk mass we may have either a cool or a warm/hot stellar disk. In the former case, the dynamics of the stars and that of the gas present are well coupled (see figure 4.11) and a grand design (or multiple-armed spiral structure if the gas content is large) is expected. In contrast, for warmer or hot stellar disks, the stars are practically inactive and the spiral structure is flocculent (see figure 4.12).

In conclusion, we are arguing that the three physical factors mentioned above control the observed morphologies, so that the distinction between SA, SAB, and SB in the Hubble diagram depends on disk mass; the sequence according to types a, b, and c is mostly determined by gas content; and finally the transition from grand design to flocculent spirals is controlled by the "temperature" of the stellar disk.

As has been stated on several occasions, this framework can be seen as a dynamical classification scheme to "explain" the observed morphologies, or conversely, as a useful tool ("dynamical window") to diagnose intrinsic properties of galaxy disks that are less easily obtained

(a)

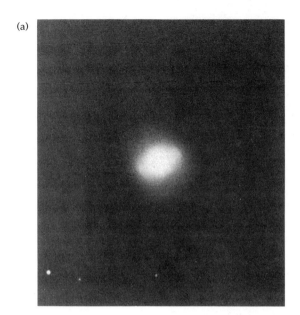

(b)

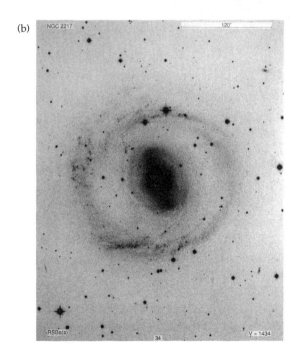

Figure 4.10
Barred galaxies [11]. A sequence of galaxies that might exemplify the case of warm and heavy stellar disks, in the order of increasing gas content: (a) NGC 4262(SB0) (b) NGC 2217(SBa).

(c)

(d)

Figure 4.10 *(continued)*
(c) NGC 4394(SBb) (d) NGC 6951(SBb/Sb).

(a)

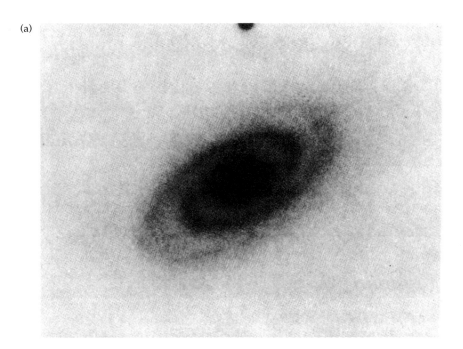

(b)

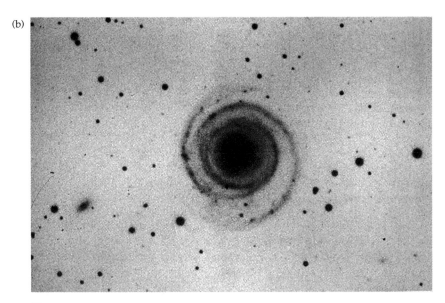

Figure 4.11
Grand design galaxies. A sequence of galaxies that might exemplify the case of cool and light stellar disks, in the order of increasing gas content: (a) a smooth-armed galaxy (Sa) [12] (b) NGC 4622 (Sb) [14].

(c)

Figure 4.11 *(continued)*
(c) NGC 3031 (Sb) from [11].

by direct observations. The main astrophysical goal being pursued here is the proper determination of the intrinsic structure of galaxies.

A common misconception frequently encountered in the literature is that some information on basic states can be obtained by a comparison of the linear growth rates of the modes with "observed" situations (stability analysis). We have shown above how the analysis of linear growth rates can only be used at the very beginning of the study to *discard* unrealistic basic states and that the real source of information on the characteristics of the basic state is the morphology of spiral structure interpreted in terms of the morphology of global spiral modes. All the global spiral modes, within the linear theory, should be moderately unstable, with comparable growth rates.

4.8   Tides and Well-Developed Bars

The general framework provided by the modal theory shows that "law and order" among the large variety of galaxy types, including the occurrence of barred morphologies, can be understood in terms of intrinsic mechanisms without resorting to mechanisms based on the action of external objects (i.e., encounters with other galaxies). This does not

(d)

Figure 4.11 *(continued)*
(d) NGC 628 (Sc) [11].

mean that the theory denies the existence of tidal effects, but rather that they are considered to be of secondary importance. Tidal forcing, when present, acts on systems that have their own global spiral modes; the resulting interaction is an interesting problem that should be studied in detail.

In contrast, it has been pointed out that in a small sample of well-known grand design spiral galaxies one often finds either that a bar is present or that a "companion" appears to be close to the galaxy of the sample. Remarks of this kind have often been taken as the ultimate rationale for the origin of grand design spiral structure in galaxies, with

Figure 4.12
Flocculent galaxies [11]. A sequence of galaxies that might exemplify the case of warm/ hot and light stellar disks in the order of increasing gas content: (a) NGC 1201(S0) (b) NGC 4594(Sa/Sb).

(c)

(d)

Figure 4.12 *(continued)*
(c) NGC 7331(Sb) (d) NGC 5055(Sb).

the corollary that grand design spiral structure cannot be generated otherwise. This conclusion transforms a possibly interesting empirical suggestion into a belief without adequate theoretical support. Indeed, we have shown *quantitatively* exactly the opposite, that grand design spiral structure, together with all other key observed morphologies, can be easily explained in terms of the properties of the intrinsic global spiral modes.

If we look more closely at the mechanism of tidal driving, we find several difficulties that make it unlikely as a general explanation for observed regular spiral structure. If this were the case, the basic states of galaxy disks should be generally stable on their own, which is hard to justify, especially if the disks are gas-rich (as is the case of M51). In addition, the required stable basic states most likely would not carry waves easily and would therefore not be bound to be easily excited from the outside. We can even imagine constructing such a delicate balance of conditions, with an ad hoc tuning of the various parameters involved, so that the disk would be ready to accept external driving but not be subject to internally self-excited spiral modes (the possibility of doing so has not been properly demonstrated so far under natural conditions); we would still have to show that good encounters are frequent. It is often recognized that galaxy encounters are not sufficiently frequent, but it is less often appreciated how "special" the encounters must be in order to have a sizable impact on spiral structure, because only *strong* and *rapid* encounters are found to be effective (as has been shown by numerical simulations). Thus we see a wide, unresolved gap between the naive empirical suggestion of tidal interaction and a quantitative demonstration that the process is a viable general explanation for regular spiral structure.

The other suggestion, that of bar driving, should be made more explicit and evaluated. Clearly, for those objects that possess only a small bar in the middle, there is not sufficient coupling between the central bar and the outer disk to claim that the outer spiral structure is driven from the inside. This can be shown quantitatively by estimating the impact of the force field generated by such a small bar at large distances. Spirals with a small bar can be actually interpreted in the opposite way (i.e., by recognizing that the small bar is created by the mode excited in the outer parts) when the disk allows for the propagation of waves close to the center. If, on the other hand, the claim is that in every galaxy there is a barely detectable broad background oval distortion of the overall field (currently being looked for in our own Galaxy), one should explain how such a broad oval distortion can drive spiral structures with the observed pitch angles and with dust lanes suggestive of a corotation circle lying in the outer parts of the optical disk. In addition, oval distortions of the disk are part of the modal analysis, so that calculations of bar-driven spirals can be put in the broader context of the generation of nonaxisymmetric structures, which includes the bars.

In concluding this chapter on modal analysis, we should mention one important and still unexplored area of interest, namely, the study of basic axisymmetric disks characterized by damped modes only. For these systems, if they exist under realistic conditions, one should investigate the spectral and spatial properties of the damped modes. Such a study should also aim at the properties of damped modes in basic states that are gas-rich (so that one may hope to model galaxies like M51). The reaction of such systems to external driving would shed quantitative light on the scenario of tidal driving as a possible explanation of regular spiral structure in some galaxies. Further discussion of possible alternative dynamical scenarios in comparison with the modal approach presented here will be given in chapter 11.

4.9 References

1. Bertin, G. 1991, in **Dynamics of Galaxies and Their Molecular Cloud Distributions**, IAU Symp. 146, ed. F. Combes and F. Casoli, Kluwer, Dordrecht, p. 93.

2. Bertin, G., Lin, C.C., Lowe, S.A., and Thurstans, R.P. 1989a, *Astrophys. J.*, **338**, 78.

3. Bertin, G., Lin, C.C., Lowe, S.A., and Thurstans, R.P. 1989b, *Astrophys. J.*, **338**, 104.

4. Bertin, G., and Radicati, L.A. 1976, *Astrophys. J.*, **206**, 815.

5. Chandrasekhar, S. 1960, **Ellipsoidal Figures of Equilibrium**, Chicago University Press, Chicago.

6. Coppi, B., and Detragiache, P. 1993, *Annals. Phys.*, **225**, 59.

7. Kelvin, Lord 1880, *Nature*, **23**, 45.

8. Mark, J.W-K. 1976, *Astrophys. J.*, **205**, 363.

9. Ostriker, J.P., and Peebles, P.J.E. 1973, *Astrophys. J.*, **186**, 467.

10. Roberts, W.W. 1992, *Annals N.Y. Acad. Sciences*, **675**, 93.

11. Sandage, A. 1961, **The Hubble Atlas of Galaxies**, Carnegie Institution of Washington, Washington, DC.

12. Sandage, A. 1983, in **Internal Kinematics and Dynamics of Galaxies**, IAU Symp. 100, ed. E. Athanassoula, Reidel, Dordrecht, p. 367.

13. Sandage, A., and Tammann, G.A. 1987, **A Revised Shapley-Ames Catalog of Bright Galaxies (RSA)**, Publication 635, Carnegie Institution of Washington, Washington, DC.

14. Strom, S.E., and Strom, K.M. 1978, in **Structure and Properties of Nearby Galaxies**, IAU Symp. 77, ed. E.M. Berkhuijsen and R. Wielebinski, Reidel, Dordrecht, p. 69.

15. Toomre, A. 1981, in **The Structure and Evolution of Normal Galaxies**, ed. S.M. Fall and D. Lynden-Bell, Cambridge University Press, Cambridge, p. 111.

16. Wevers, B.M.R.H. 1984, Ph.D. diss., University of Groningen.

II   Observational Studies

5   External Galaxies

In this and in the following chapter we shall briefly review a few observational studies of special interest. There are at least two different ways of looking at these studies. One can try to find direct empirical evidence for (or against) the presence of density waves and of global spiral modes. In other words, one can check whether the theory is well founded and consistent with the observations. Alternatively, one can *assume* that density waves or global spiral modes are present and show what can be done on the basis of the density wave interpretation of observed structures. This *application* of the theory leads to a better determination of the properties of the basic state of the galaxies under investigation; at the very least, the theory has served as a stimulus for many specific observational projects and significantly advanced our understanding of how galaxies are structured.

From the very beginning, the focus of the monograph has been on grand design spiral structure as the most challenging problem, given the existence of differential rotation. Grand design spiral structure has been sometimes downplayed in the past by arguing that only a statistical minority of objects are characterized by a grand design. Current developments in the infrared imaging of galaxies show that grand design structure in the underlying stellar disk is much more frequent than what appears from optical images.

The list of relevant observational studies given below is by no means complete. It only serves to provide readers with a general feeling for a vast research area actively in progress.

5.1   In Search of a Unifying Framework for Observations

The issue of comparing the theory of density waves with observations is very often put in terms of a confrontation, with the goal of identifying a specific observational test that could prove or disprove the theory

as a whole or in part. In particular, the core of the matter is often reduced to asking for the "ultimate measurement" that, for a *given object*
like M51, could yield a decisive answer to at least one of the following
questions: Is the observed large-scale spiral structure quasi-stationary
or rapidly evolving? Is the grand design traceable to a single global
mode? Is it driven by the companion or did the structure essentially
preexist, with subsequent minor modification by the companion?

As we stressed earlier (see section 1.4), questions of this type are only
a small subset of the questions raised by observations; although they
naturally come to mind, they do not lend themselves to a simple observational test just because they refer to the *long-term evolution* of a
complex dynamical system (over several hundred million years) far beyond our direct experience. We should be aware that similar questions
remain unresolved, at least at the level of devising a "crucial experi-
ment," even in the case of more accessible collective systems such as
laboratory plasmas or geophysical fluids. Theories predicting the long-
term evolution of complex dynamical systems cannot simply be cast
in the same form of experimental verification as elementary theories.
Nowadays, the development of very powerful computers has given
the illusion that actually predicting such long-term evolution may be
within our reach. In reality, the simulations give the time evolution of
highly idealized models; it remains to be proved to what extent these
models are representative of the dynamics of actual physical systems.
The key step is thus the modeling process itself (see chapter 7).

We therefore recognize that a simple, quantitative, decisive test of
the modal theory is not likely to be found. In other words, as we shall
show in this and in the following chapter, while the data appear to be
clearly in favor of the existence of density waves, no clear-cut answers
to the questions posed at the beginning of this section (with the example of M51) have been given, in spite of the tremendous advances in
telescopes and instrumentation in recent decades.

The modal theory presented in this monograph goes beyond this
apparent stumbling block. First of all, it provides the necessary internal dynamical justification and physical basis to the presence of density waves in galaxies. Thus we propose an internally coherent theory,
showing that a modal interpretation of spiral structure in galaxies as an
intrinsic phenomenon provides an astrophysically viable general explanation. In this respect, even if we had found a way to perform an
observational test of the kind mentioned at the beginning of this section, such a test would not be crucial for the theory as a whole. The

theory is more interested in explaining the behavior of the statistical majority of spiral galaxies.

As a result, the actual observational test for the modal theory lies in its ability to provide a consistent unifying framework for the interpretation of a large number of observational facts. Here we may repeat some of the questions raised by observations and listed earlier (in section 1.4): Why are certain galaxies barred and others not? Why are regular spiral structures generally two-armed? How can we explain the observed coexistence of morphologies within the same galaxy, especially in relation to observations in different wave bands? How do we explain the different degrees of regularity in the observed spiral structure? We have seen in chapter 4 that our theory provides answers to these questions.

As a dynamical theory, the modal theory makes statements regarding the underlying gravitational potential in spiral galaxies. Thus the focus of interest is the internal structure of galaxies. Among the various observational windows available to probe galactic structure, the one that appears to give us the best look at the mass distribution in the disk is the infrared window around 2μ. Images in this window convincingly show that the structure of the underlying gravitational potential in spiral galaxies is indeed very smooth and regular, even when the optical appearance (dominated by the distribution of Population I material) is not so regular.

5.2 Morphology

Even if, strictly speaking, this is no proof, the most convincing evidence for the existence of intrinsic global modes in galaxies comes directly from the images of beautiful grand design spirals (see figure 5.1). Anyone who cannot see the presence of global modes acting here "has no soul" (borrowing from R. Feynman's remark [13] on the role of gravity in stellar systems). Still, we would like to quantify this statement and to be more specific on some of the consequences of the modal interpretation directly inspired by the observations. A number of interesting points can be made in the morphological context.

5.2.1 Photometric Studies in Different Colors and the Existence of Density Waves

Since the time of Zwicky [44], it has been clear that the morphology of spiral structure changes considerably when a given galaxy is observed

(a)

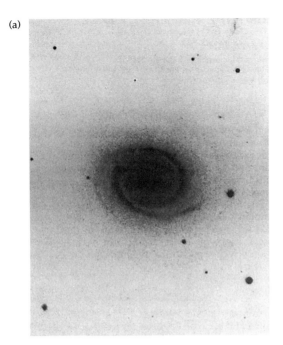

(b)

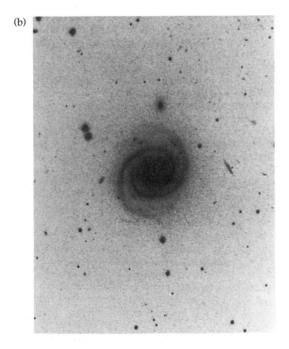

Figure 5.1
Grand design galaxies. Many grand design galaxies have been illustrated previously in the monograph. Here we show four more interesting cases: (a) NGC 1357 (Sa) [19] (b) NGC 7096 (Sa) [19].

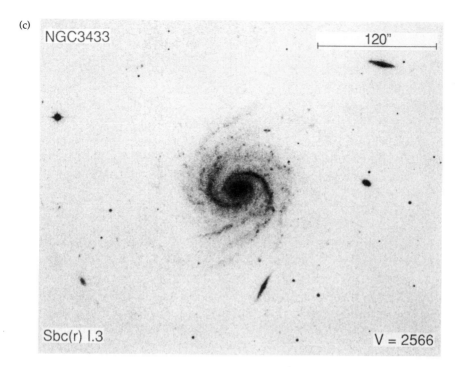

(c)

NGC3433

120"

Sbc(r) I.3

V = 2566

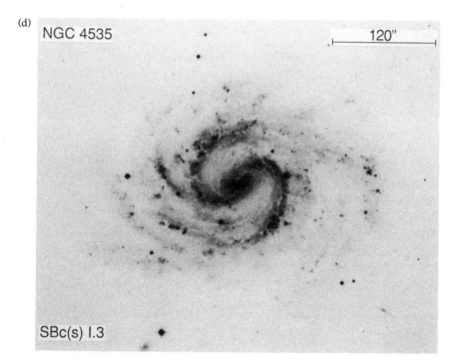

(d)

NGC 4535

120"

SBc(s) I.3

Figure 5.1 *(continued)*
(c) NGC 3433 (Sbc) [35] (d) NGC 4535 (SBc) [35].

in different wave bands. As noted earlier in this monograph (see especially chapters 1 and 3), the observed changes reflect the different dynamical behavior of the Population I disk (which is more prominent in the blue or in the ultraviolet) and of the Population II disk (which is better revealed in the red or in the infrared). More than three decades ago Oort [29] noted: "We know from the 21-cm observations in the galactic system, as well as from numerous data on other systems, that the interstellar gas is concentrated in the arms. But we do not know with any certainty whether or not *stars* contribute in an important measure to the mass of the arms. This, however, is one of the few questions concerning spiral structure which could be answered by observations, viz., by photometry of spiral galaxies in different colors." At the time this statement was made, the observational situation was not clear on this point, leaving open as a possibility that spiral arms would be mostly gaseous (i.e., Population I) features. If this were the case, an explanation of spiral structure might have involved gravity playing only a secondary role; thus one might have pursued theories arguing a magnetic origin of spiral structure (see [42]) or stochastic propagation of star formation (possibly induced by supernova explosions; see [14]). Instead, the observational studies that followed Oort's "prophecy" showed that large-scale spiral structure is indeed a density wave phenomenon.

A first indication had come from the pioneering study by Schweizer [36], which demonstrated that spiral structure in grand design objects persisted, with smoother and more sinusoidal azimuthal profiles, in the redder filters he was using. In contrast, the azimuthal profiles in the bluer filters showed sharper features, more typical of a shocked gas (see chapter 3). Note that this study came roughly ten years *after* the density wave theory had been formulated, so that we can really refer to it as an experimental verification of a theoretical prediction. The general trend was confirmed by later studies, particularly by D. Elmegreen's beautiful atlas of galaxies in two "colors" [12]. One may object that the red wave band used in this atlas (wavelengths shorter than 1μ) is significantly influenced by the Population I disk; still, it is amazing to find how smooth and regular grand design spiral structure becomes in these infrared images with respect to the more standard optical pictures. One really sees (e.g., in the atlas's two pictures of M51) how the sometimes squarish and ragged structure of the optical arms, often interspersed with spurs and other local features, becomes "distilled" into a smooth bisymmetric structure in the infrared, much like the smooth gravitational potential that is anticipated if the structure is indeed associated with global spiral modes.

Finally, in the last few years it has become possible to probe the large-scale structure of galaxies in the infrared at wavelengths longer than 1μ. At 2μ (the K-band and K'-band are at 2.2μ and 2.1μ, respectively) the observed emission is largely dominated by the underlying evolved stellar Population II disk (in particular by emission from K and M giants), which is supposed to carry most of the mass of the disk. Images in this wave band indicate the presence, in general, of low-m, astonishingly regular, and smooth density perturbations over the axisymmetric disk. Thus the evolved stellar disk is participating coherently in the large-scale spiral structure. The amplitude of the density enhancements can be very high (about 50%), but the overall smoothness and regularity and the similarity with the shapes of theoretically calculated global modes leave little doubt that the linear theory of spiral modes is probably adequate, to some extent even at the quantitative level. Thus we can claim, at least for large-scale spiral structure, that Oort's prophecy has been fulfilled and that the issue is observationally settled in favor of the participation of the old stellar disk.

5.2.2 Correspondence between Observed Morphological Types and Theoretically Predicted Global Modes

A general property to be checked for comparison with the theory is the radial extent of observed spiral structure. Grand design normal spiral structure appears to have well-developed arms out to 3–4 exponential scale lengths of the disk. Beyond the optical disk, weaker but regular spiral arms are sometimes found to continue in the cold gas (e.g., in M101, NGC 6946, and NGC 628). In the inner parts of the disk the stellar arms generally stop, especially if a bulge is present (see M81 or NGC 2997), while weaker but regular arms can continue all the way to the center (see the beautiful case of M51 [43]). This observational situation is consistent with the structure of normal spiral modes, which require the participation of a relatively cool stellar disk but also the major contribution of the cold interstellar gas. Thus their radial structure would in general set corotation in the outer disk. Because Lindblad resonances are partially transparent in the gas, the continuation of spiral arms either in the outer or in the inner parts of the disk, if gas-rich, is anticipated.

In contrast, barred modes are expected to be mostly supported by the stellar Population II disk, with corotation just outside the tip of the bar at 1–2 exponential scale lengths of the disk. Thus barred structure should be frequently prominent in red and infrared images, as observed. The size of big bars fits in with the expectations of the modal theory.

Another way to look at the issue of the size of spiral modes is to ask how the observations can lead to a detection of resonances in spiral galaxies, somewhat in the same spirit as in the case of planetary rings. Any positive determination of a Lindblad resonance or of corotation in a galaxy would automatically identify the presence of a well-defined pattern speed, which is a distinctive feature of global spiral modes. Several tracers have been invoked for the purpose, like rings (in the Population I disk) to trace Lindblad resonances, or the location of dust lanes with respect to the spiral arms (to set constraints on the location of the corotation circle; see chapter 3), and other "kinematical" tracers that will be briefly mentioned in section 5.4.

Another interesting morphological feature that finds a natural explanation in the modal context is the amplitude modulation, often observed along the spiral arms. In the modal theory, amplitude modulation reflects the presence of wave interference in the process of mode maintenance (see chapter 10). Amplitude modulations had been noted by Schweizer [36] in his study mentioned above, and it was realized immediately that those features might be used in order to locate the corotation circle of the observed spiral structure. More recently, the general concept has been pursued in great detail through careful modeling of M81, where the location and the structure of the amplitude modulation (see figure 5.2) are taken to be the major constraints for matching the theoretically predicted mode with the observed spiral structure. This type of amplitude modulation shows up prominently, and exactly as expected in the modal theory, also for barred galaxies like NGC 1300 (see figure 5.3). It also shows up clearly in some objects observed in the infrared at 2.1μ (see color plate 4 for NGC 4622), proving that this is indeed a feature in the underlying gravitational potential. In general, the relative phase between the observed maxima and minima along the arms should be different, depending on whether the interfering waves are short and long trailing or leading and trailing (see chapter 10). We should note that an ad hoc construction of a transient wave packet with a given amplitude modulation can in prin-

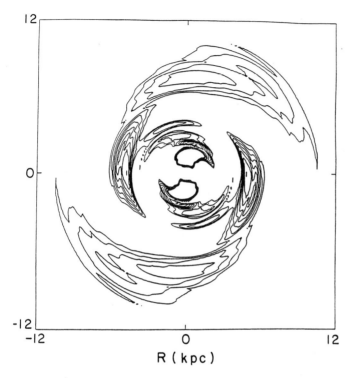

Figure 5.2
Observed amplitude modulations in M81 [11]; the contours of the $m = 2$ component are shown rectified to a face-on aspect of the galaxy.

ciple match any observed structure, although it is extraordinary to find that these observed features are so naturally explained in the modal context.

An additional morphological aspect for comparison between modal theory and observations is the structure of large bars (see also [9]). Bar modes tend to come in the form of two blobs, usually followed by a pair of trailing arms departing abruptly just outside the two lumps (see chapter 4). This two-blob structure is frequently observed. We can refer not only to the SB0 galaxies mentioned earlier, like NGC 2859 (see figure 0.7), but even to "straight bars" where the bulge is less important, as in NGC 1300. When the bisymmetric structure is distilled out of the data (see figure 5.3), the two-lump structure appears, amazingly similar to the structure of the theoretical linear bar modes. This two-blob structure is found to be even more common and more prominent in the

(a)

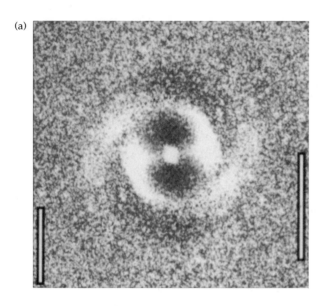

(b)

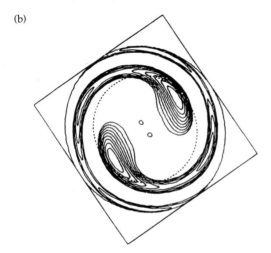

Figure 5.3
Amplitude modulations in NGC 1300 (upper frame [10]), for the two-fold symmetric part
of the galaxy rectified to a face-on aspect; amplitude modulations in a bar mode (lower
frame [3]).

infrared observations at 2.1μ, which probe the structure of the old stellar disk (see color plate 5).

Finally, we should also mention the presence of offset dust lanes inside the bar, which are taken as evidence of shocks resulting in the gas because of the presence of the more slowly rotating bar (see chapter 3). In the absence of the bar, the dust lanes tend to continue toward the center in the gas disk in some gas-rich normal spiral galaxies (see M51 and NGC 2997).

5.2.3 Number of Arms

Grand design spiral structure is generally bisymmetric. We have argued that the reason for the preference of such bisymmetric structure is to be found in the role of inner Lindblad resonance (see chapter 4). On the other hand, multiple-armed spiral structure is often recognized (see M101, M33), with long arms on the large scale. Some specific studies [18] and surveys [15,16] have aimed at quantifying the relative weights of the various m-components in spiral galaxies. Modes with $m \geq 3$ are thought to be generally suppressed in a stellar disk, while such suppression is less efficient in a gaseous disk, for which the absorption at inner Lindblad resonance (ILR) is only partial (see chapter 10). The recent studies in K- and K'-bands show that multiple-armed structure is generally absent in the old stellar disk, consistent with this expectation. We should emphasize that such expectation derives from the *modal theory*: transient density wave packets with high m might be imagined to exist and to propagate in the stellar disk, but apparently they are not observed. This statement is dramatically demonstrated by the case of NGC 309 (see figure 5.4; see also the case of M33 in plate 6). Thus it appears that $m \geq 3$ features are avoided in the old stellar disk, even if they may frequently occur in the (optical) Population I disk (see the $m = 3$ cases shown in [10]).

The same argument would lead to the conclusion that $m = 1$ structures should be ubiquitous simply because, for $m = 1$ modes, ILR is expected to be generally absent. In fact, the coexistence of $m = 1$ components was suggested by earlier studies of nearby galaxies, like M31 (during IAU Symposium 77, in Bonn, Lin made this point after the report by W. Shane [37] of a distinct kinematical asymmetry in M31). Galaxies are often lopsided [2], which can be seen as the nonlinear result of an $m = 1$ mode of the bar type. Infrared studies show that indeed $m = 1$ components are ubiquitous in the old stellar disk, not

(a)

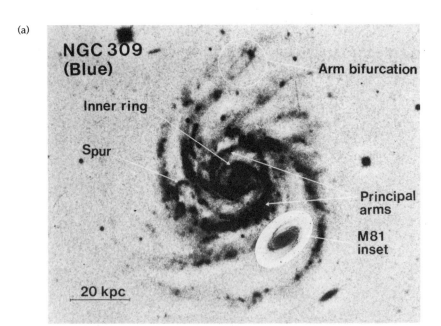

(b)

Figure 5.4
Optical and infrared appearance of NGC 309 [5]. The infrared image (b) reveals a smooth, bisymmetric structure with a prominent bar, while the optical picture (a) is characterized by multiarmed morphology.

only in the case of lopsided galaxies (such as NGC 1637, see figure 5.5 and plate 7) but also in galaxies with two well-developed arms (like NGC 2997, see figure 5.6 and plate 8).

5.2.4 Coexistence of Different Morphologies

The coexistence of different morphologies has been stressed earlier as a general theme in this monograph. In this respect, we should refer not only to the coexistence of large-scale, quasi-stationary spiral structure with small-scale spiral activity, but also to the possible development of different global modes with distinct characteristics within the same galaxy. This may occur in *spatially* decoupled regions (see, for example, the dual spiral structure of NGC 4314 of figure 5.7 or the case of NGC 4622, described in plate 4) or in *dynamically* decoupled subsystems of the galaxy, namely, the Population I and the Population II components. Because one common gravitational potential eventually affects all the various components, there would be, in general, interaction between the subsystems, so that coexistence of very different morphologies is not expected as a general rule, but only under special circumstances.

One clear case of this type of coexistence is the galaxy NGC 309, which is essentially a normal grand design/multiarmed spiral in the optical, but definitely barred in the infrared (figure 5.4). This seems to be one example where global modes have developed separately in the Population I disk and in the Population II disk, to a large extent decoupled from each other.

In the modal theory the existence of a bar is an indication that the old stellar disk is relatively heavy (see chapter 4). If an underlying bar happened to be always present in the old stellar disk, one should conclude that galaxy disks are not far, in their mass, from the *maximum disk ansatz* proposed in the context of disk-halo decomposition of rotation curves (see chapter 1). Instead, it seems that several objects are characterized by a relatively light disk. Here we may refer even to cases like NGC 2997, which may have a small oval distortion but does not display a large-scale bar, but especially to objects like NGC 4622, which is characterized by extremely thin, tight, and smooth spiral arms. This latter galaxy disk is almost certainly embedded in a heavy halo. In chapter 6 we shall see that the Milky Way is also characterized by a relatively light disk.

(a)

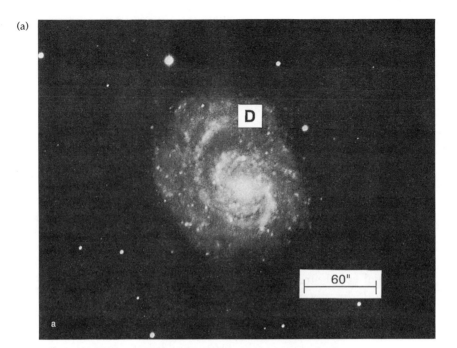

(b)

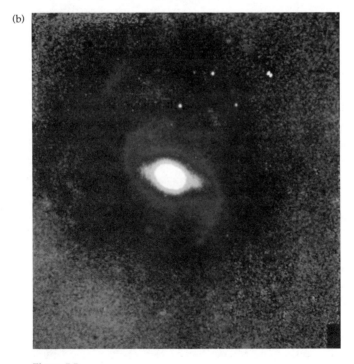

Figure 5.5
Optical and infrared (K') appearance of the galaxy NGC 1637 [4]. The lopsidedness persists in the infrared (b) in the old stellar disk, which also reveals the presence of a bar.

(a)

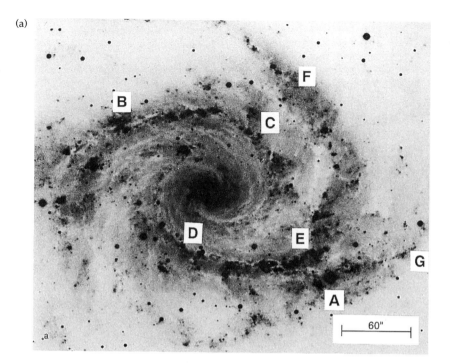

(b)

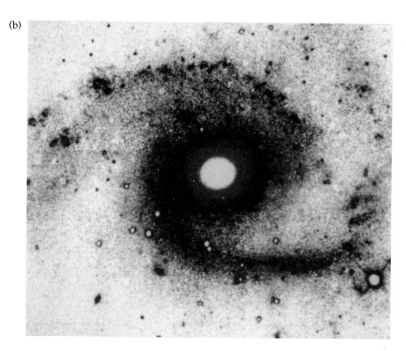

Figure 5.6
Optical and infrared (K') appearance of NGC 2997 [4]. Note that one arm (F) turns out to be mostly a Population I structure (i.e., it is essentially absent in the infrared). The infrared picture (b) clearly shows the presence of an underlying $m = 1$ component because the southern arm is definitely more pronounced.

Figure 5.7
The inner spiral in NGC 4314 is an example of dual spiral structure [34] (cf. figure 1.14).

Infrared imaging is only at its beginning. We anticipate that a study of the frequency of the underlying bar morphology in spiral galaxies will allow us to make interesting general statements on the amount and distribution of dark matter in spiral galaxies.

5.2.5   General Collective Behavior

In the early 1970s some observational studies emphasized that large-scale spiral structure is a genuine *collective* phenomenon that sees the cooperation of all the components of the galaxy under the influence of the spiral gravitational field. In particular, it was remarkable to find that the radio continuum emission from synchrotron radiation, outlining the large-scale magnetic configuration of M51, did follow the overall morphology of the grand design spiral structure ([26]; see figure 5.8).

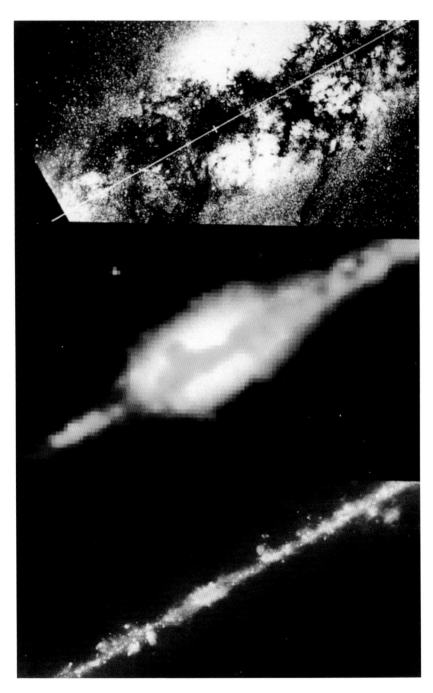

Plate 1
Three views of the Milky Way: optical (top), near-infrared (COBE, middle; cf. figure 1.4), and far-infrared (IRAS, bottom) (see chapter 1 [14]).

Plate 2
Arcs in the distant cluster Abell 2218 (see chapter 1 [26]).

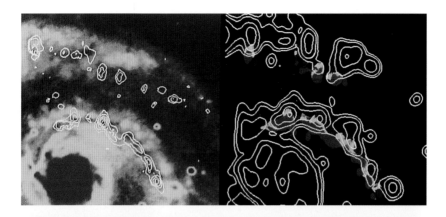

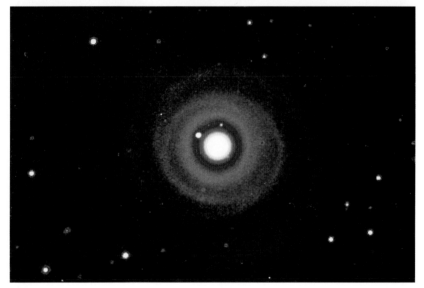

Plate 3
CO spiral arms in M51 (see chapter 3 [13]). On the left, contours representing the intensity of CO emission are superposed on an infrared image of the galaxy M51, showing that dense molecular clouds coincide with dust lanes in the spiral arms. On the right, the colors represent the line-of-sight velocity of the CO emission relative to the velocity of the galaxy (red means motion toward us, blue away) and show a sharp change in velocity as the gas moves across the spiral arm. The contours are of hydrogen emission, associated with young stars; stars seem to be forming just downstream of gas that has passed through the density wave.

Plate 4
Infrared (K') image of NGC 4622 (see chapter 5 [4]). Note the amplitude modulations along the arms. This galaxy presents dual spiral structure, with an outer pair of arms and a single inner arm winding in the opposite, presumably leading, direction. The emission in this wave band is dominated by the evolved stellar disk (K and M giants).

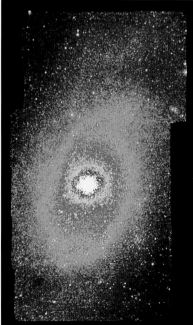

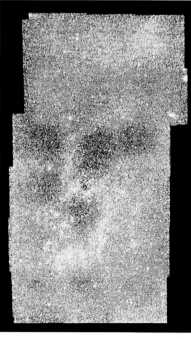

◀ **Plate 5**
Infrared (K′) image of the bar structure in the underlying stellar disk of NGC 521 (see chapter 5 [4]).

◀ **Plate 6**
The inner structure of M33 observed in the infrared (see chapter 5 [30]). The right frame displays the residual, after subtraction of an axisymmetric basic state. The morphology is dominated by bisymmetry, in contrast with the optical picture (see figure 1.12).

▲ **Plate 7**
Infrared (K′) appearance of the galaxy NGC 1637 (see chapter 5 [4]). The lopsidedness persists in the infrared in the old stellar disk, which also reveals the presence of a bar.

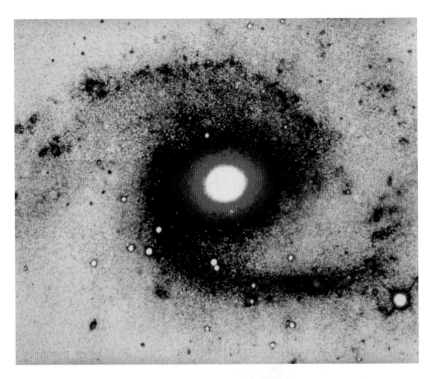

Plate 8
Infrared (K′) appearance of NGC 2997 (see chapter 5 [4]). The infrared picture clearly shows the presence of an underlying $m = 1$ component because the southern arm is definitely more pronounced. A third arm noted in the optical is essentially absent in the infrared (i.e., it turns out to be mostly a Population I structure).

Figure 5.8
The synchrotron emission in M51, measured with the radiotelescope WSRT [26].

5.3 Luminosity Classification and Star Formation

The density wave theory was at first largely inspired by the existence of orderly star formation processes along spiral arms in grand design galaxies, where HII regions and young stars could be seen like beads on a string. Thus much of the attention in the late 1960s and early 1970s was devoted to comparing theory and observations on this specific aspect of the problem. We shall report in section 5.4 on a more quantitative test of the shock scenario (see chapter 3) in the context of some detailed kinematical studies. Here we would like to record a few more general points.

Roberts, Roberts, and Shu [31] considered a sample of twenty-four galaxies, some not very regular in their large-scale spiral structure. They applied the density wave theory with no specific reference to global spiral modes and tried to fit the observed structure by means of the "local" dispersion relation for density waves (see chapter 2); curiously, they noted that fits were more satisfactory if the Q-parameter was allowed to increase above unity in the inner regions. In general, they found it appropriate to choose a pattern frequency so that corotation occurs in the outer parts of the visible disk. Some of these general remarks happen to be qualitatively in agreement with the concepts of the modal theory of normal spiral structure developed later (see chapter 4). Having therefore set empirically the main characteristics of the density waves in each object, they proceeded to evaluate the predicted strength of the shocks induced in the cold interstellar medium, on the basis of the scenario outlined in chapter 3. They then found a general correlation between the predicted properties of the shocked gas and the luminosity class of each object, as given by the classification of van den Bergh on the basis of the luminosity and arm structure of the galaxy (see section 1.3.2). This kind of "comparison" appeared to be quite satisfactory, although we should admit that the modeling was very crude, especially in view of later developments in modal theory and subsequent realization of the subtleties involved in the modeling process (see section 5.5 and chapter 7).

As we mentioned in chapter 3, a debate followed on whether there is empirical evidence that density waves actually *trigger* star formation (thus enhancing the overall star formation rate) or simply reorganize star formation processes along the arms (with a mere redistribution of the star formation events within the disk). This was the focus of several observational studies of individual galaxies (see [22]). Such studies lead to a better understanding of the physics of the interstellar medium and only indirectly to a better appreciation of the impact of density waves. We should keep in mind that a number of features, sometimes even on the intermediate scale, such as the long arm departing from one of the two major arms of NGC 2997 (see figure 5.6), are primarily features in the Population I disk, with only a minor stellar density counterpart. Thus, in evaluating the impact of large-scale spiral structure on the overall star formation rate, we should take into account that star formation processes may occur *independently* of density waves.

Evidence for large-scale shocks sweeping through the interstellar medium, of the type anticipated as a result of density waves, has been gathered in several studies, some of which are mentioned in chapter 3 [1, 22, 40]. On the very large scale, the similarity of the structure of HII regions on opposite sides with respect to the galaxy center is a sign of remarkable coherence of some star formation processes (see the ultraviolet observations of NGC 628 [8]) in areas sometimes separated by a distance of about 100 kpc (in UGC 2885, see [32] and [33]).

A somewhat different point of view gives lower priority to large-scale density waves or global modes and attempts to relate star formation (as diagnosed, for example, by $H\alpha$ emission) to spiral *activity* in general terms. It seems quite intuitive that such spiral activity would be best related dynamically to some kind of Q-parameter referred to the gaseous component. Kennicutt [20] has pursued this line of "comparison" or of "interpretation," which essentially invokes a process of self-regulation in the interstellar medium. Some correlations found are encouraging, but the study is very complex because large uncertainties are involved in the interpretation of the data (in the conversion of $H\alpha$ fluxes into star formation rates, in the estimate of the amount of molecular gas present, and in the use of average quantities for the description of the highly inhomogeneous cold gas material). In addition, as we stressed in chapter 3, for nonflocculent galaxies one expects active participation of the stars, so that a proper description of the local Jeans instability level and of the overall self-regulation process should take into account the significant contribution of the stellar disk. Although this is a difficult problem, observational studies in these directions are bound to give important observational constraints for a proper development of the density wave theory; in turn, some of the observed correlations may find a simple explanation in the overall dynamical scenario proposed by the theory. A crucial aspect of the problem that one hopes will soon be resolved is a correct determination of the amount and distribution of molecular gas across the disk.

In conclusion, although they should be hardly considered as direct evidence for the presence of global spiral modes, the observational studies of star formation processes in external galaxies have often been motivated by and are generally found to be consistent with the expectations of the density wave theory.

5.4 Kinematics

Kinematical studies offer the best opportunities for a quantitative comparison of the theory with the observations. A powerful new radio telescope (the Westerbork Synthesis Radio Telescope), commissioned in 1970, was immediately recognized to be a tool for a landmark test of the density wave theory. The most impressive work in this direction was done by Visser [39], who compared the expectations of the density wave theory with the observed kinematics of the atomic hydrogen as measured via the 21 cm line. The test consisted of several steps. First, the spiral pattern of M81 was fitted by using the short wave branch of the "local" dispersion relation, with Ω_p, the pattern frequency, considered to be a fitting parameter. (No physical explanation was given on why the system preferred the value of Ω_p that was eventually determined empirically; this is in contrast with a recent test of the modal theory mentioned in section 5.5.) Once the shape of spiral arms had been determined in this way, an estimate of the amplitude of the density wave was then made on the basis of the photometric study of Schweizer [36]. Thus the stellar data eventually gave a specification of the rigidly rotating gravitational potential associated with the spiral density wave. At this point, Visser proceeded to calculate, in the presence of the spiral field, the response of the gas in order to establish quantitatively its nonaxisymmetric distribution and especially its noncircular motions. In this step, for simplicity, the effects due to the self-gravity of the gas were not included. Finally, the predicted density distribution and velocity field of the gas were projected and smoothed on the small scale for comparison with the radio data. The general agreement between *predicted* isovelocity contours (for the velocity along the line of sight) and the *observed* isovelocity contours derived from the HI observations made at Westerbork is impressive (figure 5.9). This is generally taken to be the decisive proof of the existence of large-scale, density wave–driven streaming in M81. Further, more detailed work on the kinematics of M81 was carried out later with the Very Large Array radiotelescope [17].

Figure 5.9
Comparison between predicted (symbols) and observed (full and dashed lines) isovelocity contours for the velocity along the line of sight of the cold gas in M81, superimposed
on the observed density distribution of atomic hydrogen, which is also consistent with
the flow field [39].

The kinematics of other galaxies has been studied in great detail from the point of view of a comparison with the density wave theory. In particular, we should mention UGC 2885 [7] and M51 (especially the recent study in $H\alpha$ by Vogel et al. [41]). Great attention is being paid to the determination of velocity residuals, to be associated with the presence of a large-scale density wave.

For the case of barred galaxies, among others, such as the study of M83 quoted earlier [1], we should mention the extensive investigation of the structure and kinematics of NGC 1365 by Lindblad and collaborators [23, 24].

As mentioned earlier, one general question addressed by the studies is the search for a direct determination of the relevant pattern speed, usually on the basis of the identification of some resonance. As an example, we may recall a study by Kent [21] aimed at determining the pattern speed of the bar in NGC 936 on the basis of a method devised by Tremaine and Weinberg [38]. More recently, Canzian [6] has proposed a method for identifying corotation on the basis of the overall structure of the observed velocity residuals, which has been applied to UGC 2885. Another point of interest made recently is the possible observational identification of large-scale vortices around the stable Lagrangian points in the vicinity of corotation [27] (see also [28]).

In any case, all the studies that assume the presence of a well-defined pattern frequency and find a good empirical determination of it support the picture that the spiral structure, in the objects considered, is dominated by the presence of one mode, i.e. they *de facto* support the modal theory.

5.5 Modeling

After decades since its initial formulation, the theory of spiral structure in galaxies has reached a very advanced stage of physical and quantitative analysis. Thus it is not surprising that, as with other subjects in physics or astrophysics following an initial, rapid move, we are at a point where a major step forward, to set the properties of the basic states over which spiral modes develop, would require an extremely laborious and thorough study of the modeling process.

Indeed, the development of the density *wave* theory into a *modal* theory showed that the modeling process is much more subtle than what might have initially been foreseen. It is true that the sensitivity

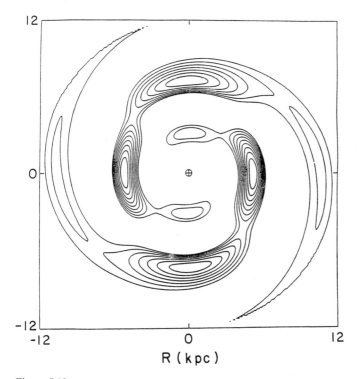

Figure 5.10
Mode calculated for a model of M81 [25]. Corotation is at about 9 kpc.

of the properties of the global modes to the characteristics of the basic state has the welcome consequence that such a modeling process, as a dynamical window to be used together with other observational windows, can help us bring the structure of galaxies sharply into focus. Still, the process requires an extraordinary amount of careful observational and theoretical work, as will be briefly outlined in chapter 7.

It is hoped that careful surveys of the physical properties of spiral galaxies will soon be able to give a more quantitative comparison with the observations for the theoretical framework of classification of spiral morphologies, based on the modal theory outlined in chapter 4. For the case of individual galaxies, a first attempt at a full modeling based on the modal theory has been made recently for the galaxy M81 [25] (see figure 5.10). The hard work required to properly implement such a modeling process will be rewarded by a major improvement in our perception of how galaxies are intrinsically structured.

158

Chapter 5

5.6 References

5.65References

1. Allen, R.J., Atherton, P.D., and Tilanus, R.P.J. 1986, *Nature*, **319**, 296.

2. Baldwin, J.E., Lynden-Bell, D., and Sancisi, R. 1980, *Monthly Notices Roy. Astron. Soc.*, **193**, 313.

3. Bertin, G., Lin, C.C., Lowe, S.A., and Thurstans, R.P. 1989, *Astrophys. J.*, **338**, 78.

4. Block, D.L., Bertin, G., Stockton, A., Grosbøl, P., Moorwood, A.F.M., and Peletier, R.F. 1994, *Astron. Astrophys.*, **288**, 365.

5. Block, D.L., and Wainscoat, R.J. 1991, *Nature*, **353**, 48.

6. Canzian, B. 1993, *Publ. Astron. Soc. Pacific*, **105**, 661.

7. Canzian, B., Allen, R.J., and Tilanus, R.P.J. 1993, *Astrophys. J.*, **406**, 457.

8. Chen, P.C., Cornett, R.H., et al. 1992, *Astrophys. J. Letters*, **395**, L41

9. Combes, F., and Elmegreen, B.G. 1993, *Astron. Astrophys.*, **271**, 391.

10. Elmegreen, B.G., Elmegreen, D., and Montenegro, L. 1992, *Astrophys. J.*, Suppl. ser., **79**, 37.

11. Elmegreen, B.G., Elmegreen, D., and Seiden, P.E. 1989, *Astrophys. J.*, **343**, 602.

12. Elmegreen, D. 1981, *Astrophys. J.*, Suppl. Ser., **47**, 229.

13. Feynman, R.P., Leighton, R.B., and Sands, M. 1963, **The Feynman Lectures on Physics**, Addison-Wesley, Reading, MA, **1**, 7-7.

14. Gerola, H., and Seiden, P.E. 1978, *Astrophys. J.*, **223**, 129.

15. Grosbøl, P. 1985, *Astron. Astrophys.*, Suppl., **60**, 261.

16. Grosbøl, P. 1988, in **Towards Understanding Galaxies at Large Redshifts**. *ASSL*, ed. R.G. Kron and A. Renzini, Kluwer, Dordrecht, **141**, 105.

17. Hine, B. 1984, Master's thesis, University of Texas.

18. Iye, M., Okamura, S., Hamabe, M., and Watanabe, M. 1982, *Astrophys. J.*, **256**, 103.

19. Kennicutt, R.C. 1981, *Astron. J.*, **86**, 1847.

20. Kennicutt, R.C. 1989, *Astrophys. J.*, **344**, 685.

21. Kent, S. 1987, *Astron. J.*, **93**, 1062.

22. Knapen, J.H., Beckman, J.E., Cepa, J., Hulst, T. van der, and Rand, R.J. 1992, *Astrophys. J. Letters*, **385**, L37.

23. Lindblad, P.O., and Jörsäter, S. 1987, *Publ. Astron. Inst. Czechoslovak Acad. Sciences*, **69**, 289.

24. Lindblad, P.O., and Lindblad, P.A.B. 1994, in **The Gaseous and Stellar Disks of the Galaxy**, ASP Conference Series, ed. I.R. King, **66**, 29.

25. Lowe, S.A., Roberts, W.W., Yang, J., Bertin, G., and Lin, C.C. 1994, *Astrophys. J.*, **427**, 184.

26. Mathewson, D.S., Kruit, P.C. van der, and Brouw, W.N. 1972, *Astron. Astrophys.*, **17**, 468.

27. Nezlin, M.V., Polyachenko, V.L., Snezhkin, E.N., Trubnikov, A. S., and Fridman, A.M. 1986, *Sov. Astron. Letters*, **12**, 213.

28. Nezlin, M.V., and Snezhkin, E.N. 1993, **Rossby Vortices, Spiral Structure, Solitons**, Springer-Verlag, Heidelberg.

29. Oort, J.H. 1962, in **Interstellar Matter in Galaxies**, ed. L. Woltjer, Benjamin, New York, p. 234.

30. Regan, M.W., and Vogel, S.N. 1994, *Astrophys. J.*, **434**, 536.

31. Roberts, W.W., Roberts, M.S., and Shu, F.H. 1975, *Astrophys. J.*, **196**, 381.

32. Roelfsema, P.R., and Allen, R.J. 1985, *Astron. Astrophys.*, **146**, 213.

33. Rubin, V.C., Ford, W.K., and Thonnard, N. 1980, *Astrophys. J.*, **238**, 471.

34. Sandage, A. 1961, **The Hubble Atlas of Galaxies**, Carnegie Institution of Washington, Washington, DC.

35. Sandage, A., and Bedke, J. 1988, **Atlas of Galaxies Useful for Measuring the Cosmological Distance Scale**, NASA SP-496, Washington, DC.

36. Schweizer, F. 1976, *Astrophys. J.*, Suppl. ser., **31**, 313.

37. Shane, W.W. 1978, in **Structure and Properties of Nearby Galaxies**, IAU Symp. 77, ed. E.M. Berkhuijsen and R. Wielebinski, Reidel, Dordrecht, p. 180.

38. Tremaine, S., and Weinberg, M.D. 1984, *Astrophys. J. Letters*, **282**, L5.

39. Visser, H.C.D. 1977, Ph.D. diss., University of Groningen.

40. Vogel, S.N., Kulkarni, S.R., and Scoville, N.Z. 1988, *Nature*, **334**, 402.

41. Vogel, S.N., Rand, R.J., Gruendl, R. A., and Teuben, P.J. 1993, *Publ. Astron. Soc. Pacific*, **105**, 666.

42. Woltjer, L. 1965, in **Galactic Structure**, ed. A. Blaauw and M. Schmidt, University of Chicago Press, Chicago, p. 531.

43. Zaritsky, D., Rix, H.W., and Rieke, M.J. 1993, *Nature*, **364**, 313.

44. Zwicky, F. 1957, **Morphological Astronomy**, Springer-Verlag, Berlin.

6   The Milky Way

Historically, studies of the Milky Way inspired and motivated the initial development of the density wave theory from the semiempirical point of view [25]. For example, they allowed one to estimate the strength of the spiral gravitational field (to show that it is only a small fraction of the axisymmetric field). They also allowed one to determine the approximate location of the corotation circle (to show that it is in the outer, gas-rich region of the galactic disk). This last point gives a strong hint that the gravitational instability of the gaseous component may play an important role. In addition, studies of the solar neighborhood, providing us with the most direct view of the basic properties of stellar populations and of the interstellar medium, have also made a strong case for the presence of a massive dark halo.

6.1 A Unique Case

The Milky Way plays a very special role in the studies of galactic structure and of galactic dynamics. On the one hand, because the Sun is located very close to the equatorial plane of the disk and in an average position, at a distance of approximately two exponential scale lengths from the center, we can measure with great accuracy several typical properties of the disk. From observations of the solar neighborhood (a region within a few hundred parsecs from the Sun), we can form the most detailed picture of the composition and of the kinematical properties of stellar populations and of the interstellar medium. Obviously, for external galaxies located at distances of 1Mpc and beyond, there is no way that we can obtain a comparable amount of astrophysical information. Indeed, for many purposes we often extrapolate to external galaxies from our knowledge of the solar neighborhood, using one typical argument of astronomy, namely, that our Sun is not likely to be in an anomalous environment.

On the other hand, just because of our location in the galactic plane, it is hard to form a global picture of the structure of our Galaxy. For geometrical reasons, it is difficult to extract from the data a reliable determination of basic kinematical quantities, such as the rotation curve, for the part of the disk outside the solar circle. Even more problematic is the determination of the global spiral structure of the Galaxy. One obstacle to this latter goal derives also from the fact that, in the plane of the disk, much of the light is obscured by interstellar dust. Thus, even if we know that our Galaxy is in the range Sb–Sc of the Hubble morphological classification and is probably neither like NGC 2841 nor like NGC 5364, we do not actually know what its appearance would be if looked at from the outside.

In conclusion, although the Milky Way is an excellent case for *local* studies, it is certainly a poor case for comparison with the theory of global spiral modes.

6.2   Spiral Arms and Spiral Features

The search for spiral arms in our Galaxy was marked by major breakthroughs in the 1950s. At first, nearby spiral arms were found to be delineated by HII regions [29]; the use of HII regions as tracers of spiral arms was pursued later (see, for example, [13]). Very soon the discovery of spiral structure in our Galaxy was confirmed and extended to a much wider spatial region by radio observations of the HI distribution (see figure 6.1; see also [9, 15]). The key observational features or "arms" are named after the constellation in whose direction they are found, like Perseus, Carina, and Orion. The Sun is closest to a feature often called the "Orion arm" or "Orion spur." Other tracers have been used to delineate spiral arms, besides optical HII regions and radio HI, among which we may mention CO clouds [10].

One line of work that historically played a major role was the attempt to place various observed spiral features in the framework of a large-scale, grand design pattern. A two-armed, large-scale pattern, calculated on the basis of the short wave branch of the local dispersion relation for density waves and of the Schmidt [33] model for the basic state, was shown by Lin and Shu ([23]; see figure 6.2) to match most of the observed spiral features known at that time. The identi-

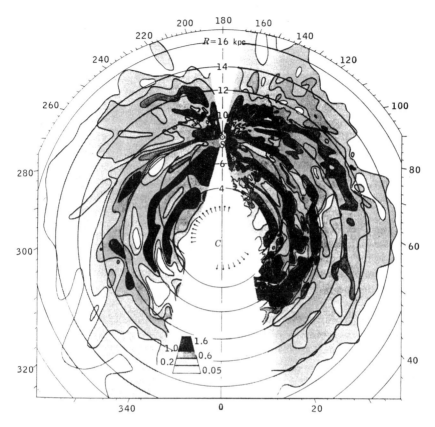

Figure 6.1
Distribution of hydrogen in the galactic plane; the density scale is in atoms/cm^3 [32, 31].

fied pattern frequency was fairly low ($\Omega_p = 11$ km sec^{-1} kpc^{-1}), setting corotation in the outer disk and ILR in the vicinity of the so-called 3 kpc arm. This two-armed pattern was admitted from the beginning to be an idealization of the complex real situation. Still it served the purpose of providing a quantitative framework for the analysis of observational data. In particular, evidence was sought (and in many cases found) for the existence of "sequential structures" of the various arm tracers (dust lanes, HII regions, young stars) in order to test the shock scenario outlined in chapter 3. Several years later, in the same spirit of aiming at the identification of *one* simple global pattern that would match all or most of the observed spiral features, it was argued that the data are better consistent with a more open, *four-armed* pattern (see figure 6.3 [13]; see also [7]). It appears that some of the conflicting results that have been obtained and the very complex observational situation argue against the actual existence of a simple, well-defined pattern over the whole disk. In other words, our Galaxy is probably *not* a grand design spiral. As we shall see in Section 6.4, some physical considerations also support this point of view.

Our Galaxy probably has a small central bar. Some kinematical studies have indicated this as a possibility. One morphological fact pointing to this has been reported recently [6]; namely, the finding of a brightness asymmetry in the infrared light distribution close to, but on opposite sides of, the galactic center, which would be naturally explained by the presence of a bar on the scale of 1 kpc, inclined with respect to our line of sight.

Radio observations at the Very Large Array telescope have revealed the presence of a minispiral very close to the galactic center [11] on the scale of only 1 pc, with ionized gas distributed in a fairly regular, three-armed pattern (figure 6.4). This is of course a phenomenon that is completely unrelated to the problem of density waves on the large scale; a hydrodynamical model for the explanation of such a minispiral has been proposed recently [12].

6.3   Kinematics

Since the early work of Kerr [17], the kinematical data derived from radio observations of atomic hydrogen showed the existence of general systematic deviations from the pure differential rotation expected in an axisymmetric disk. Some of these streaming motions were found to be associated with the presence of spiral features [8,34]. One of the early successes of the density wave theory was to show [25] that the observed "wiggles" in the kinematics of HI could be theoretically converted into density perturbations, that is, gaseous arms, of the same pitch angle, pattern frequency ($\Omega_p \simeq$ 11–13.5 km sec^{-1} kpc^{-1}), and general character as those of the pattern found earlier (see section 6.2) on the basis of other observational features. The amplitude in the gravitational field (about 5% of the axisymmetric field) associated with the underlying density wave was also found to be consistent with the amplitude inferred from the study of the migration of young stars. This latter phenomenon initially attracted much of the interest in the comparison of theory and observations. The general idea was to try to show that the birthplaces of relatively young stars could be traced back in

Spiral Pattern for 1965 Schmidt Model

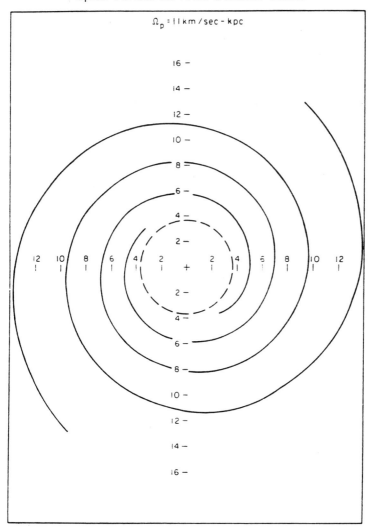

Figure 6.2
Theoretically proposed spiral pattern for our Galaxy based on the dispersion relation for density waves [23].

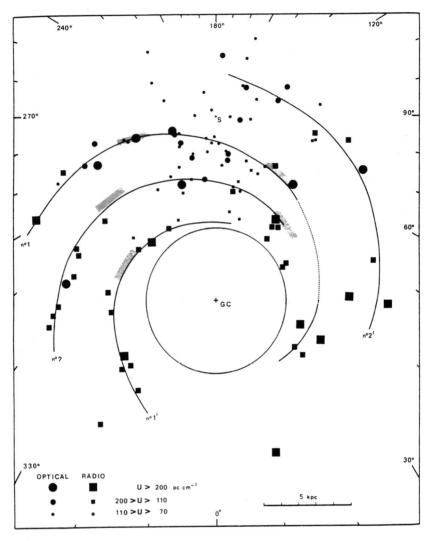

Figure 6.3
Empirically proposed four-armed pattern for our Galaxy based on a detailed study of the
location of HII regions [13] (see also [7]).

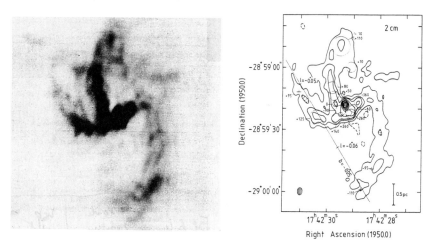

Figure 6.4
Radiograph (left) and contour map (right) for the minispiral discovered with radio observations at 2 cm at the center of our Galaxy [11]; the bar gives a scale of 0.5 pc.

time to a location inside the spiral arms, thus strengthening the shock scenario of chapter 3. This study required the calculation back in time of the orbits of a sample of stars, with known ages in the presence of a density wave of given properties (pitch angle, pattern speed, and amplitude), starting from their current state as initial conditions. Within this framework, 25 main sequence stars with age ≈ 150 Myr were studied by Yuan [37], 19 Cepheids by Wielen [36], and 400 early-type stars by Grosbøl [14].

We could mention here a few other detailed observational tests that were part of these early investigations. The detailed modeling of the HI line profiles in various observing directions was performed by Yuan [38]; the density wave was expected to produce specific deformations of the line profile for the 21 cm emission line that were to be compared with the observed shapes of the 21 cm lines. The presence of a density wave would also change the statistical distribution of stellar orbits in the solar vicinity, and this was compared with the observed deviation of the "vertex" of the velocity dispersion ellipsoid [27,39] and with the determination of the Oort constants [24].

It was immediately recognized that a systematic asymmetry, of a few km/sec, between the "basic state" derived from the HI kinematical data of the Northern Hemisphere and that derived from the Southern

Hemisphere (see figure 6.5 and [17]) remained unexplained by the simple presence of a two-armed density wave and might rather be taken as an indication of an overall broad oval distortion of the galactic gravitational field. Since then, more than two decades of observational and modeling work have tried to resolve this latter point. In particular, on the basis of the kinematics of HI, Blitz and Spergel [5] argue that an overall asymmetry exists and should be explained by the presence of a broad barlike oval distortion (with quadrupole amplitude of ≈ 0.02), slowly rotating (with pattern frequency ≈ 6 km sec^{-1} kpc^{-1}) and misaligned with the small inner bar that the same authors identified in the central region of the Galaxy (see section 6.2). In turn, the small inner bar found on the basis of the disk brightness asymmetry in the direction of the central bulge is apparently in agreement with the requirements of HI and CO kinematics as observed earlier by Sinha [35] and by Liszt and Burton [26].

This general issue of triaxiality, both on the "small" and on the "large" scale, is not fully resolved. On the "small" scale, we may refer for comparison to a recent bulge model by Kent [16]. On the large scale, a systematic survey of motions of carbon stars by Schechter and collaborators appears to suggest that the gravitational field of the basic state of our Galaxy is indeed essentially axisymmetric. A similar conclusion has been reached based on reexamination of the determination of the Oort constants by Kuijken and Tremaine [21].

In conclusion, the modeling of the kinematics observed in the gas and in the stars of our Galaxy is one area of research greatly stimulated by the hope of producing immediate tests for the density wave concept. Much progress has been made since the pioneering work of the 1960s, but the global picture is, to a large extent, still unresolved.

6.4   Modeling the Solar Neighborhood

Since the early studies of Oort [30], it was suspected that the amount of matter in the solar neighborhood required to guarantee vertical equilibrium, given the observed velocity dispersion of the stars and their vertical density distribution, exceeded by a factor of 2 the amount of matter actually observed in the form of stars and gas. For a long time, this was taken to be one of the key pieces of observational evidence for the presence of "dark matter." In this case the "missing mass" was argued to be in the disk. A more detailed analysis of more recent data appeared essentially to confirm the discrepancy, that is, the existence

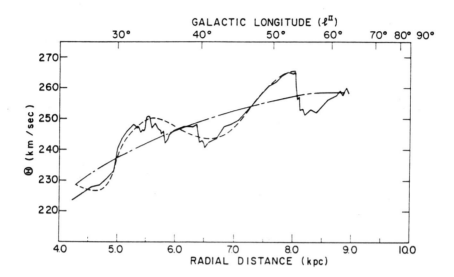

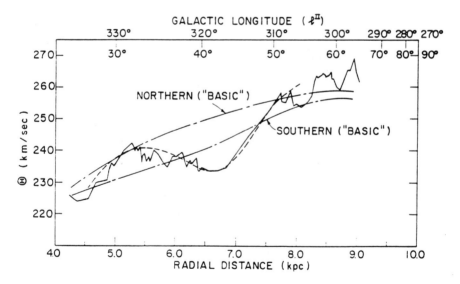

Figure 6.5
HI rotation curve for our Galaxy, Northern Hemisphere (upper frame) and Southern Hemisphere (lower frame). Observations [17] are given as solid lines; theoretical model curves based on density wave analysis are given as dashed lines [25].

of such "dark disk" component [2]. Studies of the local neighborhood served also the purpose of calibrating the scales for a global model of the basic state of our Galaxy. In one model that appeared to satisfy most of the available observational constraints [4], the disk density at the location of the Sun is approximately $\sigma_\odot \approx 63 M_\odot pc^{-2}$; this model is actually characterized by a relatively *light* disk (in the sense of our modal studies—see chapter 4), since much of the support to the rotation curve derives from the nondisk (bulge-halo) component.

In more recent years, especially through the work of Kuijken and Gilmore [19], the situation has evolved *against* the presence of a dark disk component in the solar neighborhood. In fact, a reexamination of the problem (see [3,20]) may put the value of the dynamically determined disk density down to $\sigma_\odot \approx 55 \pm 10 M_\odot pc^{-2}$, well compatible with the observed disk density $\sigma_{obs} \approx 46 \pm 6 M_\odot pc^{-2}$. The observed disk density is apparently the result of star and gas contribution, with a share $\sigma_* \approx 34 M_\odot pc^{-2}$ and $\sigma_{gas} \approx 12 M_\odot pc^{-2}$. It should be stressed that the observational basis and especially the modeling that justify these numbers and error bars are fairly complex and partly controversial. Still, the current analysis is in the direction of a possible reconciliation between observed and dynamical disk mass, with the consequence that disk dark matter is no longer needed.

Two very important points emerge from these considerations. The first point concerns the size of the dark halo in our Galaxy. Because, according to these new results, the disk is much lighter than that of the model by Bahcall and Soneira (which was already in the domain of light disks), we have to conclude that our Galaxy must be embedded in a very massive spheroidal halo. It turns out that, at the location of the Sun (i.e., at approximately two exponential scale lengths from the galactic center), the observed rotation curve is supported by a mass distribution that is 25% in the form of a bulge, 25% in the form of the disk, and 50% in the form of a dark spheroidal halo (see figure 6.6). This places our Galaxy in the domain of very light disks. Indeed, such a massive halo largely exceeds the expectations of the so-called maximum disk ansatz (see chapter 1 and the detailed study of NGC 3198 by van Albada, Bahcall, Begeman, and Sancisi [1]). Thus, from the dynamical point of view, our Galaxy should be characterized by very *tight* spiral structure. In addition, the "failure" of the maximum disk hypothesis for the case where the data are most easily available to us strongly suggests that relatively light disks may be fairly common (e.g., see the comments on NGC 4622 in chapter 5).

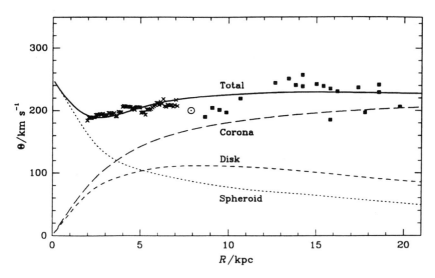

Figure 6.6
Three-component model of the Milky Way Galaxy fitted to the large-scale HI rotation curve determined in [28].

The second important point, implied by these results, is that our Galaxy is extremely gas rich, since $\sigma_{gas}/\sigma_* \approx 35\%$ already at two exponential scale lengths of the disk and this ratio is bound to increase farther out. Under these circumstances, because of the multiple-armed structure and abundant spiral activity on the small scale that are expected in the gas (see chapter 4), it is unlikely that our Galaxy possesses a simple, bisymmetric grand design.

Another interesting aspect of the problem of modeling the solar neighborhood is that of deciding, on the basis of all the available data, what the level of local stability is with respect to axisymmetric disturbances, that is, of deriving an estimate for the local effective Q-parameter. From the very beginning of the density wave theory, in the 1960s, it was realized that the combination of the effects associated with the presence of gas and of finite thickness of the disk might well make the local neighborhood very close to marginal stability (i.e., to $Q \approx 1$, for the effective Q-parameter). Still, it is unclear how to pin down the issue quantitatively. It is indeed very hard to make use of the wealth of kinematical and spatial data for various populations of stars and various gas components (each characterized by specific values of density, thickness, and velocity dispersion) as an input to our highly idealized dynamical models (either one-component,

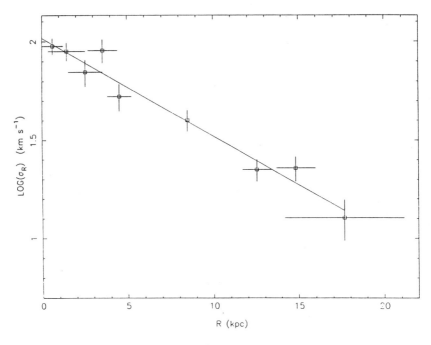

Figure 6.7
Radial profile of the radial velocity dispersion measured in the old stellar disk of our
Galaxy [22]; the fitted line implies an exponential scale length of 4.4 kpc.

zero-thickness models, or even two-component, finite-thickness models; see chapter 7 for a detailed discussion). This clearly demonstrates
that much of the basic difficulty in describing the dynamics of galaxies
lies not in the limited data available to us but rather in the *modeling* of
such complex systems. In any case, it is also clear that the value of the
effective Q—parameter so derived from the data heavily depends on
issues such as the preference of the Bahcall and Soneira model for the
mass distribution or the lighter disk solution favored by the work of
Kuijken and Gilmore.

An interesting recent observational result concerning the basic properties of our Galaxy has been the derivation of the profile of the velocity dispersion for stars representative of the disk component of the
Galaxy ([22]; see figure 6.7). Profiles of comparable quality are clearly
not available for external spiral galaxies. The monotonic, decreasing
trend found appears to be consistent with the equilibrium properties
of a constant-thickness, exponential stellar disk. It also shows that the

bulge region is relatively hot, while in the outer parts the velocity dispersion of the stars and that of the cold gas become comparable, favoring the picture outlined in chapters 3 and 4 of an outer disk where stars and gas are well coupled to each other. Here again, in the absence of better, more direct information, one is tempted to extrapolate to external galaxies the general conclusions that can be drawn for our Galaxy.

If one is not too discouraged by the difficulties of the modeling process, one might try to combine all the available kinematical and spatial data from the various components of the Galaxy and to derive a profile for an effective $Q-$parameter. For the reasons mentioned above, the role of gas on such a profile is expected to be very important or even dominant. Thus one might test the concept of self-regulation (see chapter 3). There have been claims in support of the conclusion that indeed the disk of our Galaxy is self-regulated at $Q \approx 1$ over a wide radial range [18], which is interesting but because of the complexity of the matter should be taken with caution and tested further.

6.5   References

1. Albada, T.S. van, Bahcall, J.N., Begeman, K., and Sancisi, R. 1985, *Astrophys. J.*, **295**, 305.

2. Bahcall, J.N. 1984, *Astrophys. J.*, **276**, 169.

3. Bahcall, J.N., Flynn, C., and Gould, A. 1992, *Astrophys. J.*, **389**, 234.

4. Bahcall, J.N., and Soneira, R.M. 1980, *Astrophys. J.*, Suppl. ser., **44**, 73.

5. Blitz, L., and Spergel, D.N. 1991a, *Astrophys. J.*, **370**, 205.

6. Blitz, L., and Spergel, D.N. 1991b, *Astrophys. J.*, **379**, 631.

7. Bok, B.J., and Bok, P.F. 1976, **The Milky Way**, Harvard University Press, Cambridge.

8. Burton, W.B. 1966, *Bull. Astron. Inst. Netherlands*, **18**, 247.

9. Burton, W.B. 1973, *Publ. Astron. Soc. Pacific*, **85**, 679.

10. Dame, T.M., Elmegreen, B.G., Cohen, R.S., and Thaddeus, P. 1986, *Astrophys. J.*, **305**, 892.

11. Ekers, R.D., Gorkom, J.H. van, Schwarz, U.J., and Goss, W.M. 1983, *Astron. Astrophys.*, **122**, 143.

12. Fridman, A.M., Khoruzhii, O.V., Lyakhovich, V.V., Ozernoy, L., and Blitz, L. 1994, in **The Gaseous and Stellar Disks of the Galaxy**, ASP Conference Series, ed. I.R. King, **66**, 285.

13. Georgelin, Y.M., and Georgelin, Y.P. 1976, *Astron. Astrophys.*, **49**, 57.

14. Grosbøl, P. 1976, Ph.D. diss., Copenhagen University Observatory.

15. Henderson, A.P., Jackson, P.D., and Kerr, F.J. 1982, *Astrophys. J.*, **263**, 116.

16. Kent, S.M. 1992, *Astrophys. J.*, **387**, 181.

17. Kerr, F.J. 1962, *Monthly Notices Roy. Astron. Soc.*, **123**, 327.

18. Kruit, P.C. van der 1992, in **Morphological and Physical Classification of Galaxies**, ed. G. Longo, M. Capaccioli, and G. Busarello, Kluwer, Dordrecht, p. 39.

19. Kuijken, K., and Gilmore, G. 1989, *Monthly Notices Roy. Astron. Soc.*, **239**, 605.

20. Kuijken, K., and Gilmore, G. 1991, *Astrophys. J. Letters*, **367**, L9.

21. Kuijken, K., and Tremaine, S.D. 1991, in **Dynamics of Disk Galaxies**, ed. B. Sundelius, Goteborg, p. 71.

22. Lewis, J.R., and Freeman, K. C. 1989, *Astron. J.*, **97**, 139.

23. Lin, C.C., and Shu, F.H. 1967, in **Radio Astronomy and the Galactic System**, IAU Symp. 31, ed. H. van Woerden, Academic Press, London, p. 313.

24. Lin, C.C., Yuan, C., and Roberts, W.W. 1978, *Astron. Astrophys.*, **69**, 181.

25. Lin, C.C., Yuan, C., and Shu, F.H. 1969, *Astrophys. J.*, **155**, 721.

26. Liszt, H.S., and Burton, W.B. 1980, *Astrophys. J.*, **236**, 779.

27. Mayor, M. 1970, *Astron. Astrophys.* **6**, 60.

28. Merrifield, M.R. 1992, *Astron. J.*, **103**, 1552.

29. Morgan, W.W., Sharpless, S., and Osterbrock, D.E. 1952, *Astron. J.*, **57**, 3.

30. Oort, J.H. 1932, *Bull. Astron. Inst. Netherlands*, **6**, 249.

31. Oort, J.H. 1962, in **Interstellar Matter in Galaxies**, ed. L. Woltjer, Benjamin, New York, p. 3.

32. Oort, J.H., Kerr, F.J., and Westerhout, G. 1958, *Monthly Notices Roy. Astron. Soc.*, **118**, 379.

33. Schmidt, M. 1965, in **Galactic Structure**, ed. A. Blaauw and M. Schmidt, University of Chicago Press, Chicago, p. 513.

34. Shane, W.W., and Bieger-Smith, G.P. 1966, *Bull. Astron. Inst. Netherlands*, **18**, 263.

35. Sinha, R.P. 1979, in **The Large-Scale Characteristics of the Galaxy**, IAU Symp. 84, ed. W.B. Burton, Reidel, Dordrecht, p. 341.

36. Wielen, R. 1973, *Astron. Astrophys.*, **25**, 285.

37. Yuan, C. 1969, *Astrophys. J.*, **158**, 889.

38. Yuan, C. 1970, in **The Spiral Structure of our Galaxy**, IAU Symp. 38, ed. W. Becker and G. Contopoulos, Reidel, Dordrecht, p. 391.

39. Yuan, C. 1971, *Astron. J.*, **76**, 664.

III

Dynamical Mechanisms

7

Basic Models and Relevant Parameter Regimes

A naive, widespread belief is that theoretical studies should be able to tell the dynamical evolution of a galaxy on the basis of the observational input provided by astronomers. In some cases it is recognized that observations present several limitations. Thus the picture is formed that sometime in the near future the data will be accurate enough so that we will eventually be able to deduce the dynamical evolution of a galaxy, even if at present we cannot carry out such a project. This naive point of view is fundamentally incorrect. In fact, dynamical theories address models that, even in their more complex form, greatly simplify the physics of the systems under investigation. When observational data are used as an input to these models, one has to adopt a specific modeling process, and this turns out to depend heavily on the type of phenomena that are studied. In addition, dynamical theories show that the behavior of the system (not only in terms of modes, but in more general terms) is very sensitive to the value of certain parameters, like Q. Thus we should abandon the hope for a deductive approach, and consider instead an iterative process by which dynamics serves as one additional "window," besides the various observational windows, toward the goal of determining the structure of galaxies.

7.1 Modeling of the Basic State

To be more specific, in the study of the dynamics of spiral galaxies one often considers zero-thickness, one-component models,[1] identified by three functions, the rotation curve $V(r)$, the mass density distribution

---

1. Here and below, the term *one-component* describes the modeling of the *disk*. In general, the disk is embedded in a bulge-halo spheroidal component, which is taken to be immobile.

$\sigma(r)$, and the equivalent acoustic speed $c(r)$ (see section 7.2). This idealized model is usually analyzed either in terms of fluid equations or within a stellar dynamic (kinetic) theory. A major modeling step is involved in converting observational data into constraints on these three functions, and we should stress that the process, even in the case where we had perfect data at our disposal, would depend on the particular phenomenon we are interested in. It is interesting to note that, even if in our idealized modeling we have three functions at our disposal, galaxies appear to line up in the Hubble diagram (see figure 1.1) essentially along two linear sequences (SA, SB). The modal studies try to offer an explanation for this fact.

7.1.1   Physical Galaxy as a Multicomponent Three-Dimensional System

The reason for such fundamental "ambiguity" in the modeling process is that galaxies are *intrinsically complex* systems. They are made of several components which are distributed as a fully three-dimensional system (see chapter 1). When we come down to describe the various components in a first simplified scheme, we can list the disk, the bulge, and the halo. Then, within the disk we should separate the stars from the gas. It is well known that different stellar classes have different spatial distributions and different kinematical behavior. Similarly, the gaseous disk includes different phases (hot gas, warm ionized gas, cold gas). The cold interstellar medium is made of cold gas clouds, with a very wide spectrum of masses ranging from a few solar masses up to $10^5 M_\odot$, and characterized by different kinematical properties. The ionized gas component is also under the influence of large-scale magnetic fields. All the various components that can be listed are in mutual dynamic relation because stars are continually born from the cold gas and frequently stars explode as supernovae, thus returning mass and energy to the interstellar medium. Therefore, even considering the galaxy as one isolated dynamical unit (which is certainly not true), a direct, deductive dynamical study is, strictly speaking, impossible because we should work, for such a purpose, with an unacceptably large set of equations. Note that the more accurate and complete the data are, the more evident it is that the direct, deductive approach is actually facing an impossible task.

In simpler, more constructive, terms we might ask: How should we proceed to choose the three functions $V(r)$, $\sigma(r)$, $c(r)$ on the basis of the available data and what do we lose in such a modeling process? How can we choose a less idealized but still tractable model for the purpose of studying large-scale spiral structure in galaxies? If we insist on using a zero-thickness, one-component model (because of its simplicity and flexibility), what are the missed key physical ingredients to be kept in mind so that the three functions are best chosen for our purposes?

These apparently paradoxical questions are best exemplified in the choice of the function $\sigma(r)$, in view of the finite thickness of galaxy disks, for two different goals. As a first goal, we take the construction of the mass model for a given observed rotation curve $V(r)$ (see sections 1.2.3–1.2.5). Here we would like to identify, in a first study, the functions $V_D(r)$, $V_B(r)$, $V_H(r)$, which define the disk, bulge, and halo contributions to the rotation curve, so that $V^2(r) = V_D^2(r) + V_B^2(r) + V_H^2(r)$. It is clear that in the determination of the force field associated with the disk component the effects of finite thickness of the disk are insignificant for practical purposes, so that if $\rho_D(r, z)$ represents the "actual" three-dimensional mass distribution of the disk component, we can refer to a zero-thickness disk with density distribution $\sigma(r) = \int_{-\infty}^{\infty} \rho_D(r, z)\, dz$. Finite-thickness effects can be included and are shown not to change the resulting mass models significantly. If we take instead as a primary objective the study of large-scale spiral structure for the same object, finite-thickness effects cannot be ignored. For example, for tightly wound waves in a fluid disk one can show that wave propagation is stopped altogether if the thickness of the disk z_* is sufficiently large ($z_* > \frac{2\pi G\sigma}{\kappa^2}$; where σ denotes the projected disk density from the previous example). The reason is that gravitational fields produced by a disturbance in a disk with finite thickness are "diluted" with respect to fields generated in a zero-thickness disk with the same projected disk density distribution, and such a "dilution" depends on the type of waves that are considered. For open waves, or for the purpose of constructing mass models (see previous example), the dilution can be often ignored, while for tightly wound spiral waves the dilution is very important. For this purpose, in order to make use of what can be learned on the dynamics of tightly wound waves from zero-thickness disk models, one has to resort to the concept of *active disk mass* (see section 7.2.2), which is reduced from the projected disk mass, by an amount dictated by finite-thickness effects.

From this brief discussion it should already be clear that the modeling process is indeed quite subtle. The important roles of gas in this context require a separate digression.

7.1.2   Roles of Gas

The cold interstellar medium plays a major role in the dynamics of spiral structure, as described in considerable detail in chapter 3. The different behavior of the gas (with respect to the stellar disk) at the Lindblad resonances will be described further in section 10.4; immediately after, in section 10.5, we will also elaborate on the impact of the cold dissipative gas on the nonlinear evolution of the disk especially in normal spiral galaxies, via equilibration of the spiral modes at finite amplitudes and via regulation of the disk close to the margin of axisymmetric Jeans instability. Here, in the modeling process, we should recall some specific attributes of the cold gas component.

Because of its low velocity dispersion (4–8 km/sec), the cold gas component has a very short range of action, of a few hundred parsecs, which makes it a passive component with respect to open large-scale features (such as bars). These open structures should gather their support almost exclusively from the stellar disk. In contrast, normal spiral structure, which often occurs with a relatively small pitch angle, should be modeled keeping in mind the active role of the cold gas. For practical purposes, when modeling barred spiral structure we may, in a first approximation, ignore the distribution of the cold interstellar material; for normal spirals we must take it into account.

When focusing on the problem of normal spirals, one should note that the cold gas distribution usually extends on a much larger overall scale than the stellar disk does (see figure 4.4). Thus the scale length h of the *active* disk mass is generally expected to exceed the scale h_* of the underlying exponential stellar disk. The local projected density ratio of HI to star density, σ_{HI}/σ_*, usually exceeds 15% at $r \cong 3h_*$. Typical values of σ_{HI} are a few $M_\odot/pc^2$. Because the gas is also present in the form of molecules and helium and is dynamically more active than the stars on short scales (because it is cold and is distributed in a thinner disk), the density ratio of 15% relative to atomic hydrogen shows that the gas is dynamically important, especially in the outer disk, even if at the global level the mass in nongaseous form may exceed 95% of the total mass of the galaxy.

Two-component studies of zero-thickness disks show that the impact of the colder component on local Jeans instability is indeed much larger than what is indicated by its fractional mass content. If Q_g and Q_* are the local axisymmetric stability parameters (see chapter 9) referred to the two components, then the effective stability parameter Q for the combined two-component system (see [3]) is often approximately given by $Q \approx (\frac{1}{Q_g} + \frac{1}{Q_*})^{-1}$. All the above-mentioned physical roles of the gas should be properly taken into account in the modeling process.

7.1.3 Relation between the Physical System and the Dynamical Model

In order to focus on some essential features of the general problem, we refer to an idealized situation where the system to be modeled is far simpler than the real physical galaxy to be studied, but contains a number of properties that make it far more realistic than the one-component models usually considered by the numerical codes at our disposal. Let 1(ZT) (one-component, zero-thickness) be the "model" specified by $V(r)$, $\sigma(r)$, and $c(r)$. Let 2(FT) (two-component, finite-thickness) be the "physical system" that we want to model, made of two components that we call stars and gas but we treat as fluids with finite thickness in the same force field. Then 2(FT) is specified by $V(r)$, $\sigma_*(r)$, $c_*(r)$, $z_*(r)$, $\sigma_g(r)$, $c_g(r)$, $z_g(r)$. The thicknesses are defined in such a way that the projected (column) densities are related to the peak (equatorial) densities by: $\sigma_* = 2\bar{\rho}_* z_*$ and $\sigma_g = 2\bar{\rho}_g z_g$. If each component is vertically isothermal, the volume density vertical profiles differ from the $sech^2$-solution typical of the one component isothermal slab. Still, a useful relation involving the vertical dispersion speeds can be identified:

$$(1 + \alpha)^2 = \frac{c_{z*}^2}{\pi G \sigma_* z_*} + \frac{c_{zg}^2}{\pi G \sigma_g z_g} \alpha^2, \tag{7.1}$$

where $\alpha = \sigma_g/\sigma_*$. Clearly the system 2(FT) has degrees of freedom that 1(ZT) does not possess. On the other hand, we feel that, in certain regions of the relevant parameter space and for certain purposes, we should have an adequate description of the dynamics of 2(FT) by a study of 1(ZT) if we properly choose the two profiles $\sigma(r)$ and $c(r)$.

One might naively argue that current n-body simulations *already* allow for a full dynamical study of 2(FT). However, the question of the modeling process applies at a different level to those simulations as well because we still have to make a choice for the various profiles on the basis of the available observational constraints that derive from a system much more complex than 2(FT). In addition, these n-body simulations would require the specification of the dissipative properties of the gas, to which the dynamical system is very sensitive. From the discussion given below, we will see that a proper modeling procedure is not well defined for nonlinear or time-dependent analyses.

For the purpose of modeling normal spiral structure, a procedure[2] can be devised for mapping 2(FT) into 1(ZT) based on the results of a local stability analysis (see section 9.2), if we focus on the problem of describing the properties of global spiral modes (see chapter 10). From the properties of 2(FT) the procedure determines a 1(ZT) model characterized by normal spiral modes that are also supported, with similar morphologies, in the physical 2(FT) system. It generalizes a simpler procedure developed in the context of zero thickness to incorporate the destabilizing role of cold gas in terms of "an effective Q-parameter" [3]. If we refer to open bar modes, the procedure would identify a different 1(ZT) model. Indeed, for special ranges of the parameter involved, the same 2(FT) system is expected to be able to support modes *both* of the normal and of the barred type (see chapter 10). In order to support the normal mode, the associated 1(ZT) model would have an effectively *lighter* disk than that of the 1(ZT) model associated with the bar mode. Such "mixed" or "dual" dynamical behavior is actually observed (see chapter 5).

From these considerations, a proper description of the procedure could be given only *after* the global stability problem has been addressed. The analysis would show that the modeling process is indeed very subtle. Actually, as explained below, the modeling process must go beyond these concepts.

7.1.4   Direct and Inverse Problems, in the Presence of "Imperfect" Data

The previous section assumes, for the sake of simplicity, that we have full knowledge of the seven profiles that define the two-component, finite-thickness disk. In practice, the observational data do not provide such a "sharp" picture of the physical disk. For example, from the discussion of rotation curves one finds often a factor of two uncertainty in the total disk density profile and, because of the difficulties in estimating the amount of molecules present, for a given observed value of σ_{HI} there is a similar uncertainty in the gas density profiles. Even more

---

2. Some aspects of the procedure, especially in relation to the proper choice of a softening parameter in n-body simulations, have been discussed recently in [6] (see also references therein).

uncertain are the determinations of the random motions that define c_* and, to some extent, c_g. Thus the properties of the physical system are only partly known. This brings us to consider the *inverse* problem, that is, a better determination of the properties of the physical system from the dynamical constraints imposed by the observed spiral morphology when interpreted as a manifestation of global spiral modes. Thus an iteration process should be considered, whereby our knowledge of the intrinsic structure of spiral galaxies is sharpened.

To summarize the iteration process, when focusing on an individual galaxy, the available observational data should lead to the identification of a reference model $1(ZT)_0$, on the basis of the arguments given in section 7.1.3. On the other hand, given the observational uncertainties, a whole range of $1(ZT)$ models is expected to be compatible with the observations. At this stage one can perform a modal analysis of these $1(ZT)$ models in order to identify the most plausible $1(ZT)_1$, within the allowed range, in order to match the properties of the observed spiral structure for the galaxy under investigation. In particular, the following factors should be considered: (1) the extent of spiral structure, (2) the pitch angle of spiral structure, (3) the structure of spiral arms, and (4) the overall regularity of spiral structure. Then one should consider the *inverse* problem in order to identify the available range of $2(FT)$ models compatible with $1(ZT)_1$. The iteration process would continue, because we should make sure that the $2(FT)$ model thus identified would not suffer from other undesired modes (such as open bar modes, if not observed). Guidance to the iteration process would also be provided by physical arguments, such as self-regulation (see chapter 10).

Thus we see that modeling of *individual* galaxies may become a major effort. So far, very promising results derive from the modeling of *categories* of spiral galaxies, which requires a much simpler treatment, as illustrated below.

7.2   Zero-Thickness Dynamical Models

Under the guidance of the previous discussion on the key factors involved in the modeling of galaxy disks for dynamical purposes, a family of reference zero-thickness, one-component[3] basic states has been identified. The properties of these dynamical models are now described.

---

3. See note 1 in this chapter.

7.2.1 Rotation Curve

The family of rotation curves is represented by

$$V(r) = V_\infty \frac{r}{r_\Omega} \left[1 + \left(\frac{r}{r_\Omega} \right)^2 \right]^{-1/2}. \tag{7.2}$$

It contains two dimensional parameters, a velocity scale V_∞ and a length scale r_Ω. For $r \ll r_\Omega$, $V(r)$ represents a solid body rotation with angular speed V_∞ / r_Ω, while for $r \gg r_\Omega$, the rotation curve is flat at V_∞. Note that in this discussion $V(r)$ is taken to measure the local gravitational field in the plane of the disk from the relation

$$V^2 = r \frac{d\Phi}{dr}. \tag{7.3}$$

For a "cold" disk with typical random motions $c \ll V$ the mean velocity of rotation u_θ is very close to V:

$$u_\theta = V \left(1 + O(\frac{c^2}{V^2}) \right), \tag{7.4}$$

where the correction term is usually called "asymmetric drift." Such a correction is naturally expected since pressure gradients participate with the gravitational field in the momentum balance equation. Because the correction is small in the outer disk outside of the bulge, one can argue that the function $V(r)$ can be measured directly (at least if axisymmetry is a good approximation), and indeed the family of rotation curves defined by eq. (7.2) (see figure 7.1) appears to match most of the observed curves. Thus V_∞ should be thought to be in the range of 150–350 km/sec, as observed for most spiral galaxies.

7.2.2 Active Disk Mass

For the distribution of active disk mass density we can refer to

$$\sigma(r) = \sigma_0 e^{-r/h} f + \sigma_g, \tag{7.5}$$

where σ_0 gives the magnitude of the exponential disk mass distribution, σ_g represents the gas density distribution, h is a scale length, and the reduction factor f reduces the disk mass in differential form, that is, more in the center and less in the outer parts. For the purpose of surveying several models within a simple analytical framework, one form

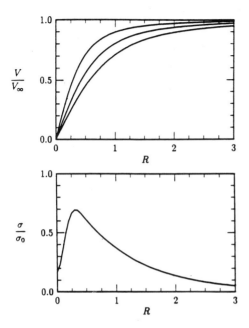

Figure 7.1
Properties of the basic states for a family of one-component models [2] (see eqs. (7.2), (7.5), and (7.6)). Quantities are plotted as a function of $R = r/h$. *Top*: Rotation curves for $r_\Omega/h = 1/2, 3/4, 1$. *Bottom*: Surface mass density for $r_{cut} = (1/2)h$ and $\sigma_g = 0$.

of f often used is

$$f = 1 - f_0\left(\frac{r}{r_{cut}}\right) + \frac{1}{6}f_0\left(\frac{r}{r_{cut}}\right)\exp(r/h),\qquad(7.6)$$

with

$$f_0(x) = (1 + 4x)(1 - x)^4 \text{ for } x \leq 1$$
$$= 0 \text{ for } x > 1.\qquad(7.7)$$

Even if this is just a simple, convenient, and smooth analytical formula, with only one new length scale (r_{cut}) introduced, the expression of f tries to take into account two different reduction effects: (1) finite-thickness effects, which are governed by the aspect ratio z_*/r, with the thickness z_* roughly constant according to the observations, and (2) possible replacement of the disk by the bulge in the central parts. The latter effect is not at all clear from the observational point of view, but it should be kept as an alternative option; in objects like M81, with a well-developed bulge, it is not clear whether the disk component actually continues in the bulge region or is actually replaced by the bulge altogether. Thus modeling of individual galaxies requires detailed explorations of cases usually not considered in a simple survey based on eqs. (7.5) and (7.6). Note also that in simple surveys σ_g is often taken to be a constant (having in mind galaxies with extended gas distributions; typical values for σ_g would be $10 M_\odot/pc^2$), while for modeling of individual objects like M51 it is more likely that the gas distribution σ_g has a scale length comparable to that of the stellar component.

Choice (7.5) often assumes that the scale length h of the exponential disk is the same as that of the observed stellar luminosity profile (typically between 1 and 10 kpc), which would be consistent with a constant mass-to-light ratio M/L for the disk, if the finite thickness corrections are all contained in the reduction factor f. Here we should mention that a constant mass-to-light ratio for the stellar disk is to be considered only as a reasonable plausible case, but is not strictly required by the observations. In addition, as noted before, the value of such M/L ratio (i.e., the magnitude of σ_0), and the magnitude of σ_g are only partly constrained by the observations (typical values for M/L are a few in solar units for the blue band and for σ_0 a few hundred $M_\odot/pc^2$).

Thus we see that observations leave wide leverage on the choice of the active disk density profile, and that the set of cases sampled by eq. (7.5) is very large and representative of a variety of realistic cases (see figure 7.2). It should be stressed that this active density profile determines the important dimensionless measure of the self-gravity of the disk,

$$\epsilon_0 = \frac{\pi G \sigma}{r \kappa^2}, \qquad (7.8)$$

which plays a key role in the dynamical analysis (chapters 9 and 10). Typically, ϵ_0 is found to be close to $\frac{1}{10}$. For Saturn rings, ϵ_0 would be much smaller, of the order of 10^{-8} or less. We should also stress again that, as a result of the presence of σ_g and of the reduction factor f the effective scale length h_{eff} of the active mass can be significantly larger than the scale length h_* of the stellar luminosity profile, often with $h_{eff} \approx 2.5 h_*$ (see figure 7.2). This point should be kept in mind particularly for normal spiral modes where both the role of the gas component and the finite-thickness effects are especially important.

Finally, we should remark that the rotation curve owes only part of its support to the disk mass distribution. Therefore, once a specific combination of $V(r)$ and $\sigma(r)$ is picked for the rotation curve and active disk mass distribution, it is usually implied that the rotation curve is supported partly by the actual disk mass $\sigma_D(r)$ from which $\sigma(r)$ is

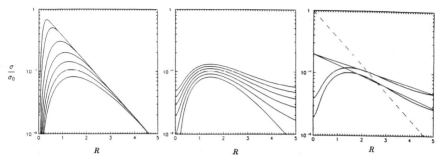

Figure 7.2
Variations on the active mass density based on a given exponential disk [2]. *Left*: The
density σ as a function of radius $R = r/h$ for $r_{cut}/h = 1/2$ (top), 1,2,3,4,5,6 (bottom).
Middle: For a case with $r_{cut}/h = 6$, the effect of the presence of the "gas component":
from $\sigma_g = 0$ (bottom) to $\sigma_g = 0.05\sigma_0$ (top). *Right*: The figure illustrates how the scale
of the active disk, because of gas and geometry, can be significantly larger than the
scale length of the optical disk: the dashed line represents the "basic" exponential disk,
the two curves ($r_{cut} = 6h$) correspond to $\sigma_g = 0.02\sigma_0$ (lower) and $\sigma_g = 0.04\sigma_0$ (upper),
while straight lines represent exponential distributions with scale 2.5 h (lower) and 3.5 h
(upper).

derived and partly by a bulge density distribution $\rho_B(r)$ and by a dark
halo $\rho_H(r)$. The latter contribution is necessary, especially for the support of the rotation curve in the outer parts, where we expect $\rho_H \sim r^{-2}$.
Thus $V(r)$ and $\sigma(r)$ cannot be chosen arbitrarily; the adopted choice
has to be checked to lead to a realistic disk-bulge-halo decomposition.
On the other hand, because the family of curves $V(r)$, $\sigma(r)$ has been
constructed for the purpose of studying large-scale spiral structure, we
should check that the implied distributions are realistic only in the radial range where spiral structure is observed (typically in the range
from $1h_*$ to $4h_*$); for example, one should not accept as factual that $V(r)$
is constant at very large radii, nor its detailed shape in the innermost
regions, if these are bulge-dominated and free from spiral structure.

7.2.3 Equivalent Acoustic Speed and Q-Profiles

Finally we come to the specification of the profile that is *least* constrained by the observations and probably has the *largest impact* on the
dynamics and evolution of spiral structure in the disk. This is the function $c = c(r)$. It is customary to specify instead the profile of the parameter $Q = c\kappa/\pi G\sigma$ because of its direct dynamical meaning (see chapter 9). The family of reference zero-thickness, one-component models
assumes the following form for the Q-profile:

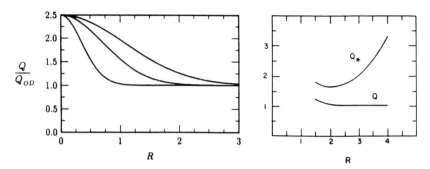

Figure 7.3
Q-profiles for a family of one-component models (see eq. (7.9)). *Left*: Q-profiles for $q =$ 1.5 and $r_Q/h = 1/2, 1, 3/2$. *Right*: The Q-profile for a specific model ($Q_{OD} = 1, r_Q = h, r_{cut} = 6h$) in the survey of [2]. Note that because of the presence of gas the effective Q-parameter can be very close to unity, as in the case shown here, even if the stellar disk is very hot.

$$Q = Q_{OD}\left[1 + qe^{-(r/r_Q)^2}\right], \tag{7.9}$$

which depends on three positive constants: Q_{OD}, q, r_Q (see figure 7.3). The parameter Q_{OD} represents the value of Q in the outer disk, that is, at radii in the range 2–4h_*. The scale length r_Q is often chosen to be of the order of h_* and is often taken to represent the scale length where the bulge is important. However, we should keep in mind that the rising of Q in the inner parts is expected even in the absence of the bulge, just because at small radii the velocity dispersion is expected to be larger (as is observed in our own Galaxy—see figure 6.7) and because the active disk mass density is expected to be low; the only exception to this trend, at least to some extent, is expected in those galaxies that are very rich in gas and are essentially bulge-free (such as M51). The parameter q is often taken to be of the order of unity.

As to the flat behavior of the Q-profile in the outer parts, this is argued on the basis of the process of self-regulation, which is expected to take place especially for the case where normal spiral structure is involved (see section 10.5 and chapter 3). Clearly, the gas tends to play a role mostly in the outer disk, where it can enforce self-regulation. In the inner disk self-regulation cannot in general be guaranteed, which gives additional physical justification to the choice of a Q-profile increasing inward. For the bar modes the exact choice of Q is less important (see chapter 10). The three-dimensional and multicomponent character of the disk is essential in the determination of the Q-profile. The dynamical behavior of the disk is very sensitive to Q, and this is why physical arguments must take precedence in its specification; otherwise, we run the risk of exploring a large variety of models that are physically not viable.

When a choice of $V(r)$, $\sigma(r)$, and $Q(r)$ is made, one should check that the numbers for $c(r)$ implied by the specified $Q(r)$ are not unreasonable. For example, it would be desirable that the profile $c(r)$ monotonically decrease with radius. Whenever values of c happened to be close to V, the concept of a thin disk would be questionable, and one should replace the disk by a bulge altogether. On the other hand, especially in the outer disk, one should make sure that $c(r)$ does not drop below the values of 5–10 km/sec that are applicable to the cold interstellar medium. Again, checks of this kind are particularly important when modeling individual galaxies.

7.2.4 Fluid Model versus Stellar Dynamics versus Fluid Disk

The above-described reference family of basic states $(V(r),\ \sigma(r),\ c(r))$ can be studied in at least three different ways. The point of view we will develop is that of a *fluid model* (see chapters 9 and 10, especially section 10.3.1), where these functions are inserted as an input to a linear stability analysis in terms of fluid equations that keeps the bulge and the halo as immobile background components. There is no claim that the real disk is fluid, nor that the system of equations is meant to describe the fluid component of a galaxy disk. The model aims at describing the real galaxy disk, which is neither a fluid nor a simple stellar system, and uses fluid equations only for their simplicity. It supplements heavily the fluid equations with physical arguments known to be applicable to the physical system being modeled. More of these qualifications will be given in chapters 9 and 10.

Another possibility is to consider the equations of *stellar dynamics* and to use $(V(r),\ \sigma(r),\ c(r))$ as an input for a linear stability analysis to this latter set of equations. A stellar dynamical disk would be described by the collisionless Boltzmann equation for the distribution function F in the four-dimensional phase space (see, e.g., [5]). Within such a stellar-dynamical treatment the Q-profile should actually be redefined, with $c/\pi \to c_r/3.36$, where c_r represents the *radial velocity dispersion* of the basic state [7]. The stellar equilibrium state is actually characterized by an anisotropic velocity dispersion, with $c_r/c_\theta = 2\Omega/\kappa$ expected in the epicyclic limit. Many of the calculations applicable or relevant to the density wave theory are indeed available within such a stellar dynamical context (see [1] and references therein). From the physical point of view, the risk of emphasizing this description is to overlook the important roles of the gas component, which instead, especially for normal spiral modes, must be taken into account (see chapter 3). Note also that the continuum description inherent in the collisionless Boltzmann equation is also an idealization of the stellar disk. Paradoxically, n-body simulations that are often set up to describe the stellar dynamics of the disk turn out to possess a fluidlike behavior due to their limited resolution in phase space (see section 10.4).

A final point of view considers the fluid equations literally as representative of a truly fluid disk. In this approach the one-component, zero-thickness fluid model makes very little sense for applications to galaxy disks, where the contribution of the stars is obviously important and their behavior is different from that of a fluid. Besides, in such a case one should seriously deal with the definition of the relevant equation of state (both for the basic state and for the density wave disturbances), which contains the physical information about the various microscopic processes that characterize the fluid disk. A hybrid version of the second and third points of view just described considers the *fluid limit* of the stellar dynamical equations (e.g., see [4] and references therein), namely, the fluidlike equations for density waves that can be derived from the collisionless Boltzmann equation. Such a fluid limit must involve an anisotropic pressure tensor. Again we should stress here that the analysis of chapters 9 and 10 should be taken in terms of a fluid model and not as a fluid limit of stellar dynamical equations.

7.2.5 Analogy with Ionized Gases

The similarities between the law of gravity and Coulomb's electrostatic interaction and between Coriolis forces in rotating systems and Lorentz forces on charged particles in the presence of a magnetic field are a source of interesting analogies between the collective behavior in self-gravitating systems and that of electromagnetic plasmas. These analogies are deeply rooted in the various methods of investigations involved and generate a natural and rewarding cross-fertilization between the two research areas. We shall only briefly and occasionally mention some of the analogies that can be drawn (see, e.g., chapter 9 for a relation with plasma oscillations), in order not to distract the reader from the primary goals of the present monograph. The plasma physicist may easily recognize a number of these points of contact, and indeed the modeling stage is particularly rich in this respect. The use of kinetic (Vlasov) equations, or magnetohydrodynamic equations, or the use of two-fluids (electrons and ions), the fluid limit of kinetic equations, the explorations of parameter space in stability analyses, and so on could be listed as part of a rich "dictionary" that would serve the purpose of promoting an exchange of ideas and methods between the two fields. In some respects, the dynamics of self-gravitating disks are similar to those of low-β plasmas ($\beta = 8\pi n T/B^2$ is the ratio of thermal to magnetic energy).

7.2.6 Scaling

The reference family of basic states identified so far might leave the impression that the following dynamical studies of chapters 9 and 10 can be automatically scaled to apply to very small galaxies and to very large galaxies. This is only partly correct, and because of this, readers should take the arguments given as applicable to typical galaxies but should always double-check when considering systems of significantly different sizes.

Probably the easiest way to get to appreciate this point is to take a look at the *timescales* that are involved. Here we list some key reference timescales and give an estimate applicable to a "typical" galaxy, starting from the shortest timescale.

A timescale τ_ℓ for a light signal to cross the galaxy disk can be 10^5 years. The lifetime of the most massive and brilliant stars (OB) can be 10^6–10^7 years. Within the same range are the period of vertical oscillation in the inner disk for an orbiting star and the dissipation time associated with the cold interstellar medium, whereas the radial and azimuthal timescales $2\pi/\kappa$ and $2\pi/\Omega$ give the dynamical timescale usually considered, with $\tau_d \approx 10^8$ years. The timescale involved in shock dissipation is probably of the same order. Larger than this, but usually smaller than 10^9 years, is the *group propagation time* (see chapter 9), defined as the time required by a density wave packet to cross a significant fraction of the disk. The secular timescale, where long-term evolution of the basic state is expected, is of the order of 10^9 years or larger, and galaxies are expected to be as old as the Hubble time, of the order of 10^{10} years. The relaxation time for star-star collision in a galaxy ($\tau_R \sim c^3/(Gm_*)^2 n$, where m_* is a typical stellar mass and n is the local number density of stars) is much longer than that, easily of the order of

10^{12} years, and this is why the stellar component of the disk, if treated separately, should be studied in terms of the collisionless Boltzmann equation.

Much of the present monograph is aimed at systems for which this *ordering* of timescales is approximately correct. However, it is clear that for very small or very large galaxies some of the ordering relations may be violated. When this happens, some of the arguments given in this monograph may not apply and may have to be reformulated. Once more, for the purpose of modeling *individual* objects, the various physical relations should be checked in quantitative detail.

7.3   Kinematics of a Differentially Rotating Disk

Much of the following discussion on dynamical mechanisms related to density waves in galaxy disks depends on the kinematical properties of the axisymmetric basic state. Because the orbits define the basic characteristics of some of the equations involved, this is true even in the continuum description of the disk in terms of fluid equations or of the collisionless Boltzmann equation.

A star orbit on the equatorial plane of such an axisymmetric gravitational field is determined by the conservation of energy

$$E = \frac{1}{2}\left(p_r^2 + \frac{J^2}{r^2}\right) + \Phi(r) \tag{7.10}$$

and angular momentum

$$J = r^2\dot{\theta}, \tag{7.11}$$

where, for simplicity, we refer to stars of unit mass, so that $p_r = \dot{r}$. Here (r, θ) are polar cylindrical coordinates. Due to the distributed mass, the potential $\Phi(r)$ is not Keplerian. If we define

$$\Omega^2(r) = \frac{1}{r}\frac{d\Phi}{dr}, \tag{7.12}$$

we find that circular orbits at a location r_0 are characterized by angular momentum

$$J = r_0^2\Omega(r_0). \tag{7.13}$$

If the angular momentum distribution is monotonically increasing with radius, so that

$$\frac{d}{dr}\left(r^4\Omega^2(r)\right) > 0, \tag{7.14}$$

then circular orbits are stable because the effective potential

$$\Phi_{\text{eff}} = \frac{J^2}{2r^2} + \Phi(r) \tag{7.15}$$

is found to possess a minimum at $r = r_0$. In general, for bound orbits with energy larger than the minimum energy of circular orbits,

$$E > E_0 = \frac{1}{2}r_0^2\Omega^2(r_0) + \Phi(r_0), \tag{7.16}$$

the radial motion is periodic and confined by two turning points r_{in}, r_{out} that depend on the integrals of the motion E and J. The associated radial frequency of oscillation Ω_r for a given potential Φ is a function of E and J. A treatment in terms of action-angle variables can be developed and is often useful. For the purpose of describing the orbits in a cool galaxy disk, the epicyclic approximation is generally clearer from the physical point of view.

7.3.1 Epicycles

The epicyclic limit of small radial oscillations applies to orbits characterized by

$$\left|\frac{E - E_0}{E_0}\right| \ll 1. \tag{7.17}$$

These orbits can be usefully separated in terms of a *guiding-center* motion (a circular orbit of radius $r_0(J)$) and of *epicyclic oscillations* (similar to Larmor gyrations of charged particles in a magnetic field). Because our galaxy models are cold, that is,

$$\frac{c^2}{V^2} \ll 1, \tag{7.18}$$

this epicyclic description is expected to hold for the majority of stellar orbits. Indeed, the typical radial excursion should be small, as measured by the epicyclic parameter

$$\epsilon = \frac{c}{r\kappa}. \tag{7.19}$$

The epicyclic frequency for radial oscillations κ (see section 2.2.1) is the small oscillation limit of Ω_r:

$$\Omega_r^2 \to \kappa^2 = 4\Omega^2 \left[1 + \frac{1}{2} \frac{d \ln \Omega}{d \ln r} \right], \tag{7.20}$$

where the expression is taken to be evaluated at $r = r_0$. The epicyclic orbit is determined from the conservation of angular momentum and is a small ellipse around the guiding center, of aspect ratio $\kappa / 2\Omega$, elongated in the azimuthal direction because Ω is generally a monotonic, decreasing function of radius. In the outer parts of the ellipse the star rotates more slowly than the guiding center (see figure 2.3). Because, for $\epsilon \ll 1$ and a given value of J, most stars are confined into a small annulus with $r \approx r_0$, the radial coordinate can be thought to be essentially the angular momentum coordinate.

The ratio $2\Omega / \kappa$ is a kind of resonance ratio. Not only does it give the aspect ratio of the epicycle, but it can be shown to give the aspect ratio of the velocity ellipsoid in the anisotropic stellar disk (see section 7.2.4); this latter ellipsoid is elongated in the radial direction. Furthermore, the ratio determines the overall shape of the orbit. Quasi-circular orbits are *closed* only if the ratio $2\Omega / \kappa$ is a *rational* number. For example, if the potential Φ is harmonic ($\Phi \sim r^2$), then Ω is a constant and $2\Omega / \kappa = 1$. In this case, the orbits are ellipses centered at the galactic center. This situation may be applicable to the innermost part of the galactic disk. If the potential is Keplerian ($\Phi \sim -1/r$), then $\kappa = \Omega$, and the orbits are Keplerian ellipses; this situation should apply to the outermost parts of the disk, in the absence of a diffuse halo.

The values of Ω and κ in the solar vicinity are related to the Oort constants that can be measured directly from kinematical studies of the solar neighborhood (see chapter 6):

$$\left(\frac{2\Omega}{\kappa} \right)_\odot = \left(\frac{B - A}{B} \right)^{1/2} \tag{7.21}$$

$$(\kappa^2)_\odot = 4B(B - A). \tag{7.22}$$

In the solar neighborhood the frequency of vertical oscillation is found to be the fastest frequency:

$$(\omega_z)_\odot = \sqrt{4\pi G \rho_\odot} \approx 3.5\Omega_\odot, \tag{7.23}$$

with $\rho_\odot \approx 10^{-23}\mathrm{g/cm^3}$ the local mass density. Note that celestial mechanics, as a subject, usually assumes $\kappa \approx \Omega$, while in galactic dynamics a reference case is $\kappa \approx \sqrt{2}\Omega$ (applicable to *flat* rotation curves).

7.3.2   Rotating Frame

If we look at the orbits in a rotating frame, rotating at angular speed Ω_p, so that $(r, \theta) \to (r, \varphi)$, with $\dot{\varphi} = \dot{\theta} - \Omega_p$, the proper Hamiltonian is the Jacobi integral defined as

$$H = E - J\Omega_p. \tag{7.24}$$

Note that the epicyclic frequency does not change in this new frame of reference, while the *resonance ratio* is changed:

$$\frac{2\Omega}{\kappa} \to \frac{2(\Omega - \Omega_p)}{\kappa}, \tag{7.25}$$

showing that an orbit may be closed in the inertial frame, but not in the Ω_p-rotating frame, or vice versa. For a given choice of Ω_p, the condition (m is a positive integer)

$$\frac{m(\Omega - \Omega_p)}{\kappa} = \pm 1 \tag{7.26}$$

is traditionally called "Lindblad resonance."

7.3.3   Shear Flow and Basic Resonances

It is important to realize that because rotation is *differential*, the above-defined resonance conditions are met at different radii in the galactic disk. At this stage Ω_p is an arbitrary constant; however, when it assumes a dynamical basis (e.g., as the pattern frequency of a given spiral disturbance; see chapter 8), then the resonance conditions take on an important physical meaning (it is at resonances that energy can be exchanged between a wave and the basic state; see chapter 9).

The situation is conveniently summarized in a *shear diagram* (see figure 2.9 for a case with $m = 2$), where we plot the profiles of $\Omega - \kappa/m$, Ω, and $\Omega + \kappa/m$ (having mostly in mind the cases $m = 1, 2, 3$). A horizontal line drawn at Ω_p may meet these curves at different radial locations. For a given Ω_p and a given m, the intersection radii r_{ILR}, r_{co}, r_{OLR} define the inner Lindblad resonance, corotation, and outer Lindblad resonance with the conditions:

$$\Omega(r_{\text{ILR}}) - \frac{\kappa(r_{\text{ILR}})}{m} = \Omega_p \tag{7.27}$$

$$\Omega(r_{\text{co}}) = \Omega_p \tag{7.28}$$

$$\Omega(r_{\text{OLR}}) + \frac{\kappa(r_{\text{OLR}})}{m} = \Omega_p \qquad (7.29)$$

For given Ω_p, m, and rotation curve, not all the resonances necessarily occur. For example, for $\Omega_p > 0$, $m = 1$ the inner Lindblad resonance is generally absent.

7.3.4 Lindblad's Kinematical Waves

The observation that the quantity $\Omega - \kappa/2$ can be roughly constant in a wide radial range of the galaxy made B. Lindblad propose the concept of kinematical spiral waves (already described in section 2.2.2). In a frame rotating at angular speed $\Omega - \kappa/2$ (supposed to be constant) in a wide radial range, orbits would appear as closed ellipses, much like the orbits in a harmonic potential seen in the inertial frame of reference. A set of orbits initially organized so as to display orbit crowding along two spiral arms would be able to *persist* as a *kinematical spiral wave* (see figure 2.4). The winding direction and the pitch angle of the spiral arms would be *arbitrary* just because the orbits are considered as a set of independent oscillators. In practice, $\Omega - \kappa/2$ is not constant, nor can the orbits be taken as independent oscillators because self-gravity and pressure effects need to be taken into account. The linear stability analysis of chapters 9 and 10 includes these dynamical effects, which go beyond the single-star description. It is interesting to note how kinematical studies have immediately drawn attention to the possibility of density waves and to the preference of bisymmetric structures in grand design galaxies.

7.4 References

1. Bertin, G. 1980, *Physics Reports*, **61**, 1.

2. Bertin, G., Lin, C.C., Lowe, S.A., and Thurstans, R. P. 1989, *Astrophys. J.*, **338**, 78.

3. Bertin, G., and Romeo, A.B. 1988, *Astron. Astrophys.*, **195**, 105.

4. Hunter, C. 1979, *Astrophys. J.*, **227**, 73.

5. Palmer, P.L. 1994, **Stability of Collisionless Stellar Systems**, Kluwer, Dordrecht.

6. Romeo, A.B. 1994, *Astron. Astrophys.*, **286**, 799.

7. Toomre, A. 1964, *Astrophys. J.*, **139**, 1217.

8   Geometry of Wave Patterns

Much of the discussion in this monograph deals with the interpretation of observed morphologies as a result of waves and modes, often at the level of a linear theory. Some terms, such as *number of spiral arms*, *pitch angle of spiral structure*, or *barred spiral structure*, are often applied without further explanation to the observed objects and to the theoretical tools developed for their interpretation. It is the purpose of this short chapter to resolve some of the ambiguities that may arise in this context.

8.1   Density Maxima, Potential Minima, and Number of Arms

If we restrict our attention to the case of even disturbances on the galaxy disk (see section 2.3), the morphology of spiral structure is often outlined as the locus of maximum disturbance at each radius r, drawn by varying the radial coordinate (i.e., the spiral pattern), or as contours of positive departures from the basic axisymmetric distribution. Ideally, one would like to refer to the total mass density perturbation because this is directly related to the nonaxisymmetric part of the gravitational potential. In practice, such spiral morphologies would be best diagnosed in terms of infrared imaging techniques, especially in the wave band close to 2μ, that are expected to trace mostly the underlying old stellar disk distribution. These images are found to give very smooth morphologies. In contrast, most of the available pictures are in more standard optical wavebands, which tend to emphasize the gas behavior, with sharper and less regular features. In this latter case, the observed spiral features are not well representative of the underlying gravitational field, although they are, to a large extent, determined by it.

When we talk about spiral arms in a given galaxy disk, we generally
refer to the density maxima, or alternatively, to the associated gravitational potential minima, for the nonaxisymmetric part of the disk.
For specific purposes, the relevant disk population involved should be
also mentioned because the spiral structure might be different in different populations, for example, essentially two-armed in the older stellar
disk and multiple-armed in the younger Population I disk.

The *number of spiral arms* thus defined bears only some relation to the
number m of the Fourier analysis commonly used to describe density
waves. Indeed, all the various Fourier components behave independently of each other in the linear theory, so that a potential perturbation
of the form

$$\Phi_s = \Phi_s(r, \theta, t) = A(r)\cos(\omega t - m\theta + \Psi(r)) \tag{8.1}$$

is often called an "m-armed spiral disturbance." On the other hand, the
m arms in an *observed* galaxy need not be as symmetric as in eq. (8.1)
(see figure 8.1 and the K'-image of NGC 2997 in figure 5.6), thus indicating the participation of different m-components (see figure 8.2).
Another example of such broad terminology is the expression "barred
structure." From the *theoretical* point of view, one often associates the
term with an $m = 2$ mode-like disturbance, such as

$$\Phi_B = A(r)\cos\Psi(r)\cos(\omega t - 2\theta), \tag{8.2}$$

while obviously a Fourier analysis of an *observed* bar would show the
participation of many even m-components, $m = 2, 4, 6, \ldots$. Bisymmetry
is thus not synonymous with $m = 2$ structure.

8.2 Pattern Frequency

Another term often used is *pattern frequency* of the spiral structure.
When applied to an observed galaxy, this means that, in a suitably chosen frame of reference rotating at the *pattern frequency*, the spiral morphology appears essentially stationary, at least temporarily (see discussion in sections 2.4 and 4.4). The statement is generally applied to
the *large-scale, grand design* spiral structure only. It has very little to do
with a linear theory because, in producing a quasi-stationary pattern
rotating at a given pattern frequency, a few linear modes are expected
to participate and to have reached a state of equilibration at nonlinear
amplitudes.

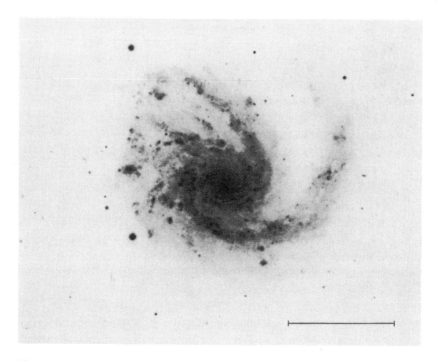

Figure 8.1
The asymmetric galaxy NGC 4254 [7].

To be sure, when a linear theory is developed, one often speaks of the "pattern frequency" associated with a linear spiral mode. In this context, the pattern frequency is defined as

$$\Omega_p = \frac{\mathrm{Re}(\omega)}{m},$$ (8.3)

where the value of ω is thought to be determined by a linear eigenvalue problem (see chapters 9 and 10 for a description of such a linear theory of a fluid model). The theorist then argues that the observed pattern frequency should be closely approximated by the value of the pattern frequency associated with the dominant spiral mode (of the linear theory).

8.3   Pitch Angle of Spiral Structure

The *pitch angle of spiral structure* at a given location is defined as the angle between the arm and the circle of radius r drawn at such radial

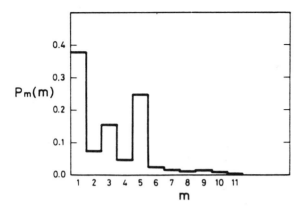

Figure 8.2
Fourier analysis of the spiral structure of NGC 4254 according to [5], revealing the dominance of odd m-components (see also the study reported in [4]). Nowadays, studies of this type can be performed on infrared images that better probe the underlying potential well in the disk.

location. A small pitch angle corresponds to tightly wound spiral structure. Empirically, when the pitch angle for a given grand design spiral is studied, it is found to stay often approximately constant with radius. In other words, observed long arms are often closely approximated by logarithmic spirals.

In the common representation of the linear theory of eq. (8.1) one generally defines a *local radial wave number* as

$$k = \frac{d\Psi}{dr} \tag{8.4}$$

so that the radial spacing between the arms is

$$\lambda = \frac{2\pi}{k}. \tag{8.5}$$

Thus, the pitch angle i of the spiral pattern described by eq. (8.1) is defined by

$$\tan i = \frac{m}{rk}. \tag{8.6}$$

We should note that the limit of tightly wound waves (see section 9.2) requires that the quantity $|rk|$ be large. The asymptotic analysis may thus yield satisfactory results even when the radial spacing λ is relatively large just because of the factor 2π involved in the relation between λ and k. For example, for $i \approx 12°$, $\lambda \approx r$, and the limit of tightly wound waves may still yield useful information!

8.4 Trailing and Leading

The winding direction of spiral arms can be either *trailing* or *leading* relative to the overall direction given by the rotation of the galaxy disk (see figure 2.8). Spiral structure is generally observed to be trailing, that is, winding opposite to the rotation for an observer moving outward along the arms. In the simple representation of eq. (8.1), trailing corresponds to $k < 0$ (for positive m and counterclockwise rotation, $\Omega > 0$), while leading is associated with $k > 0$. Note that even if the spiral arms are trailing, they may be well represented by trailing linear modes (see chapter 10) that may be composed of both trailing *and* leading waves. In particular, a barlike disturbance such as that described by eq. (8.2) might be thought of as composed of a trailing and of a leading wave. A better appreciation of the wave number composition for a given observed spiral morphology is provided by the α-spectrum (see section 10.3.2).

8.5 Normal Spiral Structure, Barred Spiral Structure, and Oval Distortions

There is a continuous range of morphological types from normal spiral structure to barred spiral structure to oval distortions of the disk. *Normal spiral structure* is the term used for those spiral morphologies that are free from the presence of a large scale bar; normal spiral structure may have a small bar in the middle if the spiral arms are found to be traced all the way to the galactic center. A small bar in the middle is thought to be dynamically unimportant for the development of large-scale normal spiral structure.

Barred spiral structure is associated with a bar feature that affects a sizable fraction of the disk. When the positive bisymmetric density contours for the perturbation are drawn for a barred galaxy (see NGC 1300 in figure 5.3), one often notes two well-defined blobs on opposite sides, which may continue outward into a pair of tightly wound spiral arms. Sometimes such a two-blob structure is directly apparent from the image of the galaxy (especially in SB0 galaxies; see NGC 2859 in figure 0.7), which includes the light from the axisymmetric basic state. Therefore, whenever the linear theory leads to the finding of $m = 2$ global modes characterized by a *two-blob structure* over a large fraction of the disk (see sections 10.2 and 10.3 for the so-called B-modes), one speaks of "bar modes" (even if real bars may have several even m-values involved). This terminology is even more natural if one thinks of the spatial structure of the bar modes that connect the axisymmetric MacLaurin sequence to the triaxial Jacobi sequence in the problem of classical ellipsoids (see figure 4.6). The case of oval distortions of the disk is associated with broad bisymmetric disturbances of the disk as a whole (without continuation into spiral arms).

Another morphological aspect related to spiral arms is that of the amplitude modulation along the arms, which refers to the possibility of finding a sequence of local maxima and local minima in the amplitude of spiral arms while moving *along* the arms. This is often observed, even in the spiral arms of barred galaxies, such as NGC 1300, and it is naturally interpreted as the result of an interference pattern between the separate waves that maintain the mode (see chapter 10).

Asymmetries of spiral arms may result from the presence of $m = 1$ spiral modes coexisting with higher m structures. Lopsidedness of the disk may be thought as a broad oval distortion of the $m = 1$ type.

8.6   Trapping of Orbits at Resonances and Self-Consistency

As a continuation to the kinematical analysis developed in the previous chapter (see section 7.3), one might consider the structure of stellar orbits in the presence of a small amplitude spiral or barred disturbance, as represented by eq. (8.1) or (8.2), that is, with a well-defined pattern speed $\Omega_p = \omega/m$. The small amplitude requirement can be expressed as the limit where the tangential forces introduced by the spiral disturbance are much smaller than the basic radial force, that is,

$$\left| \frac{mA}{r^2\Omega^2} \right| \ll 1. \tag{8.7}$$

It is clear that under these circumstances only the Jacobi integral is an integral of the motion and that the orbits are best studied in the Ω_p-rotating frame of reference; as a result, a star would change its energy and its angular momentum according to the rule

$$\dot{E} = \dot{J}\Omega_p. \tag{8.8}$$

By analogy with other well-known problems in celestial mechanics (such as the restricted three-body problem often exemplified by the Sun-Jupiter system), one can look for the location of *Lagrangian points* in such a perturbed gravitational field, that is, the location of stationary equilibrium points for stellar orbits in the Ω_p-rotating frame of reference. It is easily found that the spiral disturbance represented in eq. (8.1) admits, in general, $2m$ Lagrangian points, all located at or in the vicinity of the corotation circle. Curiously, the m points corresponding to the potential *minima* are *unstable*, while the other m points at the potential *maxima* are *stable*, so that we expect in general to have stellar orbits trapped around these latter points. The epicyclic description of stellar orbits can be usefully extended to this nonaxisymmetric case. Indeed, one finds that the guiding center orbits become distorted circles away from the corotation circle, while they become organized into m "islands" or "banana-shaped" orbits around the stable Lagrangian points, separated from the distorted circles by a separatrix running through the unstable Lagrangian points (see figure 8.3). The guiding centers move along the banana-shaped orbits in long libration periods (which scale as $A^{-1/2}$ and may be of the order of 10^9 years), while epicyclic motions occur essentially at the unperturbed value of the fast epicyclic frequency. The figure of guiding center orbits is somewhat reminiscent of magnetic islands in toroidal magnetic configurations for fusion plasmas, and indeed several analogies could be drawn between the two cases. We should also recall that the standard restricted three-body problem of celestial mechanics gives rise to five Lagrangian points, all located at or close to corotation. The two stable ones are at the tips of equilateral triangles with two vertices at the Sun and at Jupiter, and several asteroids are known to have their orbits trapped around them.

Trapping of orbits is naturally expected at resonances and traces the important process of energy exchange between a wave and the particles that make the medium. At the Lindblad resonances a different kind of trapping occurs, whereby the long axes of the ellipses (which are no longer closed because of the presence of Φ_s) librate around certain preferred stable directions. These trapping phenomena would be best described in terms of action-angle variables.

One line of research (see [6] and references therein) has pursued the investigation of these "nonlinear" stellar orbits in great detail, with the goal of identifying under which conditions a set of stellar orbits could be found able to *support* the imposed spiral field Φ_s. This is a very difficult task, as already apparent from the tendency of stars at corotation to be trapped *out-of-phase* with respect to the imposed gravitational field. It is not at all clear whether such nonlinear, stationary, self-consistent solutions are indeed available, especially because the potential Φ_s is

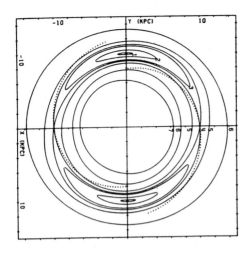

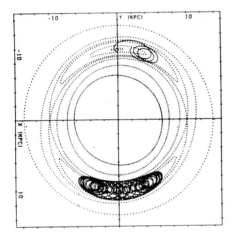

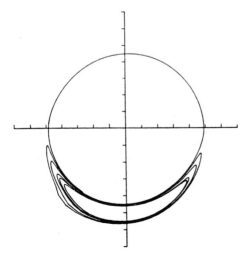

generally taken to be of arbitrary form (in spite of some similarities with the gravitational field associated with observed spiral structure). In addition, even if such an investigation were to succeed, it would remain unclear why the system has chosen the specific observed solution and not others. Furthermore, this detailed orbital description might be of interest for purely stellar disks, while large-scale normal spiral structure appears to be associated with the Population I disk, which is largely dominated by the cold interstellar medium.

In conclusion, detailed studies of stellar orbits in the presence of imposed spiral patterns may be very interesting and offer insight on various dynamical processes relevant to the problem of spiral structure, but they cannot tell the whole story. For this purpose more powerful tools have to be used, and they are found to be rooted in the theory of dispersive waves in inhomogeneous media, as will be made clear in the following two chapters.

8.7   References

1. Berman, R., and Mark, J.W.-K. 1979, *Astrophys. J.*, **231**, 388.

2. Bertin, G., Coppi, B., and Taroni, A. 1977, *Astrophys. J.*, **218**, 92.

3. Contopoulos, G. 1973, *Astrophys. J.*, **181**, 657.

4. Grosbøl, P. 1988, in **Towards Understanding Galaxies at Large Redshifts**, ed. R.G. Kron and A. Renzini, Kluwer, Dordrecht, ASSL **141**, 105.

5. Iye, M., Okamura, S., Hamabe, M., and Watanabe, M. 1982, *Astrophys. J.*, **256**, 103.

6. Patsis, P.A., Hiotelis, N., Contopoulos, G., and Grosbøl, P. 1994, *Astron. Astrophys.*, **286**, 46.

7. Sandage, A., and Tammann, G.A. 1987, **A Revised Shapley-Ames Catalog of Bright Galaxies** (RSA), Publication 635, Carnegie Institution of Washington, Washington, DC.

Figure 8.3
Trapped orbits at corotation (see also figures 4.2 and 4.3) *Top*: Structure of guiding center orbits in the presence of a two-armed spiral disturbance with the density maxima drawn as dotted curves [1]; corotation is at 10 kpc and the islands are centered at the potential *maxima*. *Middle*: Trapped star orbits [1], which include rapid epicyclic oscillations (see also [3]), are compared with the guiding center orbits. *Bottom*: Weakly trapped (guiding center) orbits [2] are easily detrapped and scattered in the presence of time-dependent perturbations.

9   Local Properties of Waves: The Dispersion Relation for the Fluid Model

From the discussion given in chapter 7 it should be clear that the physical system that we would like to study is neither a pure collisionless collection of stars nor a pure fluid. In addition to being intrinsically multicomponent, it is a truly three-dimensional system. In this chapter we describe in detail the collective properties of a zero-thickness fluid model (see section 7.2), which turns out to be able to capture the essential properties of the large-scale dynamics of spiral galaxies. Of course, such a model should not be taken literally nor used blindly, but rather considered under the guidance of what we know of the actual physical system under investigation. Most of the results and methods presented here are also available, with some modification, in the context of (collisionless) stellar dynamics (see [1] and the several papers cited there).

A rigorous derivation of the dispersion relation for tightly wound waves (see section 9.1) and of the more general cubic dispersion relation (see section 9.3) is available in the literature. Here we would like to proceed in a heuristic way so as to emphasize some important physical aspects and to move quickly to the applications, which can be easily compared with the observations and with the results of exact numerical integrations.

Many of the concepts used are well known in the general context of dispersive waves. Readers are referred to the book by Whitham [4] for a thorough introduction to the theory of dispersive waves, with examples taken from several research areas, such as the study of water waves. As is always the case in the context of dispersive waves, the dispersion relation plays a central role and is found to contain the explanation for a surprisingly wide variety of dynamical behaviors.

9.1   Heuristic Derivation of the Dispersion Relation for Tightly Wound Waves

Because gravity forces are the key physical ingredient of the problem, we should start out by describing the concept of Jeans instability. In a sense, Jeans instability has little chance to develop in a real system; rather, the scales of real systems can be interpreted as determined by Jeans instability as developed in the past.

9.1.1   Homogeneous Fluid Model

Consider a formal equilibrium configuration made of an infinite, homogeneous, self-gravitating fluid of density ρ_0 and sound speed c initially at rest. For many reasons, such equilibrium configuration is unsatisfactory, still it can be seen to represent a real system with respect to *small-scale* disturbances. Thus a discussion of the merits of such an idealized starting point can be postponed to the interpretation level.

For such a fluid, linear perturbations are associated with the dispersion relation

$$\omega^2 = k_{tot}^2 c^2 - 4\pi G \rho_0. \tag{9.1}$$

Here ω and k_{tot} denote frequency and wave number of the elementary wave. Note that the no-gravity limit $G \to 0$ yields sound waves, while the cold limit $c \to 0$ gives a "free fall" kinematic solution. The frequency

$$\omega_J = \sqrt{4\pi G \rho_0} \tag{9.2}$$

is the natural frequency associated with self-gravity. Marginal stability occurs at the *Jeans wave number*, defined by

$$k_J = \frac{\omega_J}{c}. \tag{9.3}$$

Jeans instability occurs at long wavelengths $k_{tot} < k_J$, for which the collapse tendency of gravity overwhelms pressure forces. Note that a spherical, nonrotating cloud of self-gravitating matter has a size R, which, as a result of the virial theorem, is of the order of its own Jeans length (as can be constructed in terms of average values for the density and for the velocity dispersion).

An analogy could be drawn with the case of electrostatic plasma waves. For these, the opposite sign of the Coulomb force between equal charges requires that the relevant dispersion relation be modified by replacing $-\omega_J^2 \to \omega_{pe}^2 \equiv 4\pi n e^2 / m_e$, where the plasma frequency ω_{pe} is expressed in terms of the particle density n, the electron charge e, and the electron mass m_e. In this case, the cold limit gives stable plasma oscillations ($\omega^2 \approx \omega_{pe}^2$), while the length obtained by analogy with the Jeans length is the Debye length $\lambda_D \equiv c/\omega_{pe}$, which describes the screening occurring in a plasma in the presence of an imposed static electric field.

9.1.2 Effects of Rotation

The linear stability analysis of a similarly idealized homogeneous fluid rotating at a constant angular velocity Ω yields a quadratic dispersion relation in ω^2. For $\mathbf{k}_{tot} \cdot \boldsymbol{\Omega} = 0$ the relevant dispersion relation is

$$\omega^2 = 4\Omega^2 + c^2 k_{tot}^2 - \omega_J^2, \tag{9.4}$$

which readily shows the stabilizing role of rotation. Note that the Jeans length estimated on the basis of representative values for ρ_0 and c in the solar neighborhood is of the order of 1 kpc, which correctly identifies the *vertical* scale height of the galactic disk.

9.1.3 Disk Geometry and Differential Rotation

In order to move on to the goal of describing the dispersion relation for density waves in a thin differentially rotating disk, we should include three main effects and modify eq. (9.4) accordingly. The first modification arises naturally because the desired basic state is axisymmetric and is characterized by the presence of mean flow motions (differential rotation). Thus polar cylindrical coordinates are the natural choice and the convective derivative $d/dt = \partial/\partial t + \mathbf{u}_0 \cdot \nabla$ requires a Doppler shift of the frequency $\omega \to \omega - m\Omega(r)$, where m/r is the azimuthal wave number of the perturbation. Second, in the presence of differential rotation the stabilizing contribution due to Coriolis forces is expected to become $4\Omega^2 \to \kappa^2$, where κ is the epicyclic frequency. Finally, the potential theory for (even) density waves on the thin surface of the disk modifies the destabilizing term associated with self-gravity $4\pi G\rho_0 \to 2\pi G\sigma|k_{tot}|$, where σ is the disk density; the factor 2 results from the Gauss theorem. (For odd, bending waves, the self-gravity term would have an opposite sign, corresponding to a stabilizing, restoring force; see section 2.3.)

Therefore, this heuristic discussion leads to the dispersion relation

$$(\omega - m\Omega)^2 = \kappa^2 + k^2 c^2 - 2\pi G\sigma |k|, \tag{9.5}$$

which is indeed what is recorded in section 2.3. Here we have replaced k_{tot} by k following the ordering $k_{tot}^2 = k^2 + m^2/r^2 \approx k^2$, $|rk_{tot}| \gg 1$, under which the dispersion relation can be justified by a rigorous derivation. The first requirement is the condition of tightly wound waves. The second requirement allows for a local relation between perturbed potential Φ_1 and perturbed density σ_1.

9.1.4 Kinematical Limit

The "kinematical limit" of the dispersion relation that is obtained by dropping the effects of finite velocity dispersion and of self-gravity ($c \to 0$, $G \to 0$) gives

$$\omega - m\Omega = \pm\kappa, \tag{9.6}$$

which recovers Lindblad's idea of dispersion orbits characterized by the condition $\Omega_p = \omega/m = \Omega \pm \kappa/m$ (see section 7.3).

9.2 Properties of the Dispersion Relation for Tightly Wound Waves

We have just argued that elementary density waves of the form $\sigma_1 = \tilde\sigma_1$ $\exp\left[i\left(\omega t + \int k dr - m\theta\right)\right]$ can occur in a self-gravitating disk subject to a simple dispersion relation, quadratic in $|k|$, which can be cast in dimensionless form:

$$D(\nu, |\hat{k}|) = 0 \tag{9.7}$$

with

$$D = \nu^2 - 1 - \frac{1}{4}Q^2\hat{k}^2 + |\hat{k}|. \tag{9.8}$$

Here we have introduced the parameter

$$Q = \frac{c\kappa}{\pi G\sigma} = \frac{\epsilon}{\epsilon_0}, \tag{9.9}$$

the dimensionless radial wave number

$$\hat{k} = 2kr\epsilon_0, \tag{9.10}$$

and made use of the dimensionless parameters ϵ and ϵ_0 introduced in chapter 7. The Doppler-shifted dimensionless frequency $\nu = (\omega - m\Omega)/\kappa$ was introduced in chapter 2. Recall that corotation and Lindblad resonances are characterized by $\nu = 0$ and $\nu = \pm 1$, respectively. The dispersion relation is symmetric for the separate transformations $\nu \to -\nu, k \to -k$.

9.2.1 Stability

The condition of marginal stability ($\nu^2 = 0$) is usually discussed in the $(\hat{\lambda}, Q^2)$ plane, with $\hat{\lambda} = 1/|\hat{k}|$, where a parabola separates the stable region $\nu^2 > 0$ from the unstable region $\nu^2 < 0$. The vertex of the parabola occurs at $\hat{\lambda} = 1/2$ and $Q^2 = 1$. Thus, for $Q < 1$, there is a range of wavelengths for which the system is (locally) unstable, while for $Q > 1$ the combined effects of pressure (at small wavelengths) and rotation (at large wavelengths) make the system Jeans-stable with respect to *all* values of $\hat{\lambda}$. Note that marginal stability in parameter space (i.e., the condition $Q = 1$) is obtained by setting $\nu^2 = 0$ and by looking at which values of Q the two roots of the parabola "meet" (the vertex of the parabola).

A physical argument based on the regulating role of the dissipative gas suggests that, for light disks (i.e., the low-J limit as defined in section 9.3), the outer parts of the galaxy be close to marginal stability, with $Q \approx 1$ (see section 10.5). Obviously, a disk cannot survive long with $Q < 1$ because fast local instabilities would rapidly develop and heat up the disk at least to a regime where $Q \approx 1$.

9.2.2 Wave Branches

For a given value of ν, the dispersion relation admits up to four separate wave branches. There can be either *short* or *long* waves:

$$|\hat{k}_S| = \frac{2}{Q^2}\left[1 + \sqrt{1 - (1 - \nu^2)Q^2}\right] \tag{9.11}$$

$$|\hat{k}_L| = \frac{2}{Q^2}\left[1 - \sqrt{1 - (1 - \nu^2)Q^2}\right], \tag{9.12}$$

and each type can be either *leading* ($\hat{k} > 0$) or *trailing* ($\hat{k} < 0$).

For a given model (i.e., a specified set of functions $c(r)$, $V(r)$, $\sigma(r)$), and for given values of m and of the pattern frequency Ω_p, the *propagation diagram* is a plot of the dispersion relation in the (k, r) plane or, alternatively, in the $(rk/m, \nu)$ plane. It easily reveals the four-branch structure of the dispersion relation. Note that, for Q-profiles of the type argued in chapter 7, the propagation diagram shows that the various branches can meet at several locations (see figure 9.1). In particular,

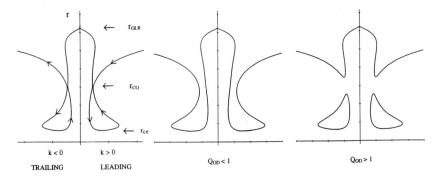

Figure 9.1
Propagation diagrams for the quadratic dispersion relation. The left frame illustrates the properties of the diagram in detail for a model that might support a tightly wound spiral mode. The radial coordinate is on the vertical axis, starting with $r = 0$. The radial wave number k is on the horizontal axis. Note the properties of the wave branches in the vicinity of the marked locations (r_{co}, r_{OLR}, and r_{ce}). Arrows denote direction of group propagation along each wave branch. The middle and right frames illustrate the qualitative change induced in the diagram by changing the value of Q_{OD} below or above the "marginal" value of 1.

for $Q_{OD} = 1$ short and long waves meet at the corotation circle r_{co}. A similar merging, close to the corotation zone, occurs at two different locations on opposite sides of r_{co} for $Q_{OD} > 1$ or does not occur at all (for $Q_{OD} < 1$). Another point where short and long waves meet is at r_{ce}, close to the general neighborhood where the Q-profile is significantly higher than Q_{OD} (the so-called Q-barrier). Another point where two wave branches formally meet (long leading and long trailing) is at a Lindblad resonance. Usually, this latter case takes place only outside, at the outer Lindblad resonance, because the inner Lindblad resonance is often "screened" by the Q-barrier.

When the argument of the square root in eqs. (9.11) and (9.12) is negative, then the wave branches exist only as complex roots and correspond to evanescent waves; these, in many cases, may be responsible for interesting tunneling effects. The examples shown in figure 9.1 illustrate the propagation diagram based on eq. (9.5) for a model that may support a tightly wound global spiral mode.

9.2.3   Wave Propagation Properties

The properties of azimuthal propagation are trivial (the wave rotates rigidly with angular speed $\Omega_p = \omega/m$). The properties of radial propagation are easily derived from the expression for the group velocity

$$c_g = -\frac{\partial \omega}{\partial k}. \tag{9.13}$$

Thus the direction of radial propagation is given by

$$\text{sgn}(c_g) = s_k s_\nu s_b, \tag{9.14}$$

where $s_k = \text{sgn}(k)$ (-1 for trailing waves), $s_\nu = \text{sgn}(\nu)$ (-1 for $r < r_{\text{co}}$), and s_b identifies the wave branch (-1 for short waves). We can assign arrows according to the various wave branches in a propagation diagram to identify the direction of group propagation.

The points where two wave branches meet in a propagation diagram are real *turning points* for wave packets, as can be seen from the fact that arrows are "turned back" (see the vicinity of r_{ce} in figure 9.1, where c_g vanishes); at these turning points wave packets undergo refraction. From this point of view, the corotation circle is also a turning point for the waves. Note, however, that, for $Q_{\text{OD}} < 1$, a wave packet would formally cross corotation with infinite group velocity.

All these are signs that, much as in other WKB problems, turning points are to be treated with special tools because, strictly speaking, the dispersion relation concept breaks down there. These different tools can take into account *wave-wave* or *wave-particle* linear interactions (wave-particle interaction can occur at r_{co} where a resonance is involved).

From this discussion we can appreciate how thoroughly the propagation diagrams give a synthetic view of the location of resonances, propagation, turning points, and even stability for density waves in galaxy disks. It is on the basis of these properties that we can understand the structure of global eigenmodes from the elementary description of a "local" dispersion relation.

9.2.4   Wave Action

Finally, from the theory of dispersive waves we recall that wave action is carried by density waves of amplitude a, with local density

$$\mathcal{A} = \left(\frac{\partial D}{\partial \omega}\right) a^2. \tag{9.15}$$

The density of wave action changes sign across the corotation circle, since $\text{sgn}(\mathcal{A}) = s_\nu$. Inside, at $r < r_{\text{co}}$ (i.e., for $\nu < 0$) the presence of a density wave rotating *slower* than the basic state lowers the angular momentum content of the system, as opposed to the more standard situation ($r > r_{co}$, $v > 0$) where the wave has positive density wave action. We recall that the angular momentum density $\mathcal{G}$ and the energy density $\mathcal{E}$ of the wave are related to $\mathcal{A}$ as

$$\mathcal{G} = m\mathcal{A} \tag{9.16}$$

$$\mathcal{E} = \omega\mathcal{A} \tag{9.17}$$

From these densities the relevant fluxes associated with a wave packet are easily derived:

$$\mathcal{F} = c_g\mathcal{A}, \tag{9.18}$$

so that $\mathrm{sgn}(\mathcal{F}) = s_k s_b$.

9.3   The Cubic Dispersion Relation

A more general dispersion relation can be derived *without* requiring the waves to be tightly wound. For this purpose, the local potential theory for open waves (for which $r^2 k^2$ is no longer supposed to be large) is based on the assumption that m^2 is large. It is possible to proceed formally and to judge later the value of the dispersion relation on the basis of its ability to interpret the solutions of the full integro-differential problem. Such a comparison has indeed been made and has shown that such an approximate description, namely, the use of the local dispersion relation, is very powerful and accurate even beyond first expectations.

If we introduce the dimensionless total wave number with magnitude $K = 2\epsilon_0 r\sqrt{k^2 + m^2/r^2}$ and assume $\epsilon_0 \ll 1$, $K = O(1)$, we can derive (see [2, 3]) the following dispersion relation for a zero-thickness, one-component, fluid model of a galaxy disk:

$$\frac{Q^2}{4} = \frac{1}{K} - \frac{(1 - v^2)}{K^2 + \frac{J^2}{(1 - v^2)}}, \tag{9.19}$$

where v, Q are defined as before and the new parameter J is defined as

$$J = 2m\epsilon_0 \left(\frac{2\Omega}{\kappa}\right) \left|\frac{d \ln \Omega}{d \ln r}\right|^{1/2}. \tag{9.20}$$

Note that, for tightly wound waves or $m = 0$ waves, K reduces to $|\hat{k}|$ (see eq. (9.10)). The symmetry properties for $v \to -v$, $k \to -k$ are preserved in this more general dispersion relation.

9.3.1 J and Q

The J-parameter is proportional to m and to the disk density σ through the parameter ϵ_0. It does not depend on the equivalent acoustic speed c. For the following discussion we would refer mostly to the value $m = 2$, which is most relevant for grand design spiral galaxies. We shall point out (section 10.4) that higher m-values are expected to be discouraged as a result of absorption at the inner Lindblad resonance. Then it is clear that J can be either of the order of unity or small depending on whether the galactic disk is largely self-gravitating or embedded in a massive bulge-halo spheroidal component. Note that, for $J \ll 1$, the dispersion relation reduces to the quadratic relation of eq. (9.7) applicable to tightly wound waves.

This more general relation is a cubic in the magnitude of the total wave number K. It can be shown that, for

$$J Q^3 > \frac{16\sqrt{2}}{27}, \tag{9.21}$$

only one real root for K is allowed, while three real roots are available for $J Q^3 < 16\sqrt{2}/27$. Of course, not all the real roots necessarily represent propagating wave branches, since for these $K \geq 2\epsilon_0 m$. In any case, for high values of J, we have at most two wave branches (a leading and a trailing open wave), while, for low values of J, we can have up to six wave branches (short, long, and open waves of either trailing or leading form). Indeed, for most applications the cubic reduces to the quadratic relation for tightly wound waves (eq. (9.7)) whenever J is below 0.5.

9.3.2   Propagation Diagrams

As suggested for the simpler regime of tightly wound waves, propagation diagrams give a direct synthetic picture of the relevant dispersion properties of density waves within a given galaxy model. The four cases shown in figure 9.2 represent the propagation diagrams for four qualitatively different basic states. Clearly, the lower left frame is recognized as a model that is approximately well described by the regime of tightly wound waves (see figure 9.1). In contrast, the top left and bottom right frames correspond to models subject to open waves only. The upper right frame has hybrid properties, with the inside in the high-J regime and the outside clearly in the low-J regime. We shall show later

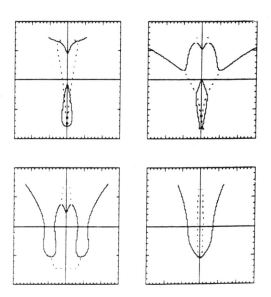

Figure 9.2
Propagation diagrams for the four mode prototypes of figure 4.5, based on the cubic dispersion relation [2]. The diagrams here use as a vertical axis the quantity v, instead of the radial coordinate, ranging from -1 (bottom) to $+1$ (top); thus the horizontal line identifies corotation. The radial wave number is given in dimensionless form on the horizontal axis $(rk/2)$ from -15 to $+15$. Dashed lines mark evanescent waves. Note that the outer Lindblad resonance ($v = +1$) is not reached by the very open wave branch with vanishing k, in contrast to the case of the quadratic dispersion relation (cf. figure 9.1).

(chapter 10) how these features have a direct counterpart in the properties of global spiral modes.

9.3.3   The (J, Q) Diagram

We can conveniently summarize the properties of the relevant parameter regimes in a (J, Q) diagram, which is actually best shown in logarithmic coordinates (see figure 9.3). The line $Q = 1$ represents the marginal stability condition for the case of tightly wound waves. Thus physical systems characterized by $J \ll 1$ are expected to be represented by points in the vicinity of such a line (region A in the diagram). We have shown previously that such a marginal stability condition is obtained by imposing that the short and the long wave branches meet at $v^2 = 0$; if two separate propagating solutions existed at $v^2 = 0$, a whole range of wavelengths would be available for which $v^2 < 0$. This distinction between locally unstable and locally stable disks has been

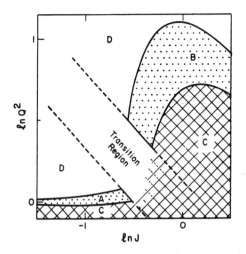

Figure 9.3
The (J,Q) diagram associated with the cubic dispersion relation [2]. The regime of moderate instability is indicated by the dotted strip, going from regime A to regime B across the transition region as the disk self-gravity is increased (see text for explanation).

illustrated by the different character of the propagation diagrams in the vicinity of the corotation circle.

By analogy, we can argue that in the regime of open waves ($J \approx 1$) the marginal stability condition is identified by requiring that at $\nu = 0$ the two available wave branches (open leading and open trailing) meet, at $k = 0$. This identifies a parabola-like curve in the (J, Q) diagram, which is displayed as a broadened strip of moderate instability (region B) applicable for the case of open waves. Close to the transition line $J Q^3 = 16\sqrt{2}/27$, local marginal stability is not so easily defined.

This discussion leaves the parameter plane divided in two parts, C and D, separated by a strip of moderate instability. Any disk with parameters falling into lower region C is expected to evolve rapidly away from it as a result of the presence of violent dynamical instabilities.

This detailed description should be confronted with more direct, numerical stability analyses in order to check whether the asymptotic theory indeed identifies the essential stability and propagation features appropriately. Such direct analyses can be performed either in terms of integrations that follow the time evolution of a given initial disturbance or in terms of the properties of truly global eigenmodes of the whole disk. In view of the applications to the problem of spiral structure in galaxies, we prefer to follow the latter route, which will be further explored in chapter 10.

9.4   References

1. Bertin, G. 1980, *Physics Reports*, **61**, 1.

2. Bertin, G., Lin, C.C., Lowe, S.A., and Thurstans, R.P. 1989, *Astrophys. J.*, **338**, 104.

3. Lau, Y. Y., and Bertin, G. 1978, *Astrophys. J.*, **226**, 508.

4. Whitham, G. B. 1974, **Linear and Nonlinear Waves**, Wiley, New York.

10

Excitation and Maintenance of Global Spiral Modes

Our discussion so far has had three basic themes. The first theme is the balance between *complexity* and *simplicity* in the problem of dispersive waves and collective behavior in inhomogeneous self-gravitating fluid disks. The complexity is readily recognized in the existence of long-range forces, resonances, and turning points; the mathematical problem is integro-differential (see section 10.3.1), the model breaks down at resonances, and the relevant boundary conditions are subtle. Still, a posteriori, the behavior of the system is remarkably simple, and the use of asymptotics is a generous source of physical insight.

The second theme is the modal approach itself. There are other ways of trying to investigate the collective behavior of density waves in galaxy disks, but it appears that the modal approach is the most natural and useful for studies of coherent large-scale phenomena; from the mathematical point of view, it is best suited for including the role played by the inhomogeneity of the system.

The third theme is the role of simple models: their limitations, and their advantages. The real system that we would like to describe is three-dimensional and made of several components; some of these would be best described by kinetic equations. A consistent analysis of the global stability of such complex system in terms of a reasonably complete set of equations would be practically impossible; in contrast, a simple fluid model can be justified and gives significant insight, provided its limitations are recognized and complemented by a thorough physical discussion.

10.1 Feedback and Amplification

Consider a second-order differential equation of the form

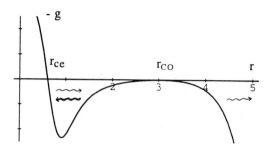

Figure 10.1
Illustration of the two–turning point problem, indicating feedback, radiation boundary condition, and overreflection. In the two simple regimes A and B identified in section 9.3.3, the integro-differential eigenvalue problem for linear modes can be approximated by an ordinary differential equation (10.1) of the type illustrated here.

$$\frac{d^2y}{dr^2} + g(r, \omega)y = 0, \tag{10.1}$$

with the function $-g(r, \omega)$ of the form sketched in figure 10.1. Here ω is a parameter that can be thought to trace a time dependence present in the original problem for an unknown function $Y(r, t) = \exp(i\omega t)y(r)$. Thus, where the equation has an oscillatory character ($g > 0$), we may imagine to identify two independent solutions as two waves with opposite propagation properties (with respect to the radial coordinate r).

When the appropriate boundary conditions are set at $r = 0$ and at $r \to \infty$, an eigenvalue problem is defined, even if the equation in general is not of the standard Sturm-Liouville form. Thus only for selected values of ω ("eigenvalues") can the boundary conditions be satisfied; the corresponding solutions are called "eigenfunctions" of the problem.

The problem corresponding to a function g of the type sketched in the figure is characterized by two *turning points* (i.e., two zeros of the function g, one at $r = r_{ce}$ being a *simple* turning point, and the other, at r_{co}, a *double* turning point). The notation ω, r_{ce}, r_{co} intentionally repeats some of the notation of chapter 9, because we intend to use eq. (10.1) as a reference equation to describe global spiral modes in a galaxy disk. We shall keep the discussion in general terms first, in order to characterize the basic dynamical ingredients that lead to the excitation and to the maintenance of global spiral modes. In section 10.2, we shall specialize the concepts developed to the two regimes for density waves introduced in chapter 9.

10.1.1 Feedback

In order to set up a global standing wave pattern in an oscillatory system, as already outlined in physical terms in chapter 4, it is necessary to have the participation of waves with opposite directions of propagation. Thus in a galaxy disk it is important to have a feedback mechanism in the central parts of the disk able to return an incoming density wave signal as an outgoing wave, back to the outer parts of the disk. Such a returned wave system is said to be able to "maintain" the global mode, that is, to support a standing wave pattern. Without such a feedback, a wave packet based on one of the wave branches of the dispersion relation would propagate away and would not be associated with a long-lasting wave pattern.

In the model equation (10.1) this feedback process is automatically present. Indeed, the profile of the function $g(r, \omega)$ at $r \sim r_{ce}$ represents a wave barrier; the turning point at r_{ce}, much like the turning point for a particle in a potential well, acts as a mirror for an incoming elementary signal, that is, for one of the two independent oscillatory solutions admitted by the equation just outside r_{ce}.

Much like in the description of the reflection (or refraction) of an electromagnetic wave in an inhomogeneous medium, the feedback process is usually supplemented by a regularity condition (at $r \to 0$), which selects for $r < r_{ce}$ the evanescent solution in the second-order differential equation. This mathematical condition is justified when no central energy sources are considered.

10.1.2 Radiation Boundary Condition

The problem defined by the function $g(r, \omega)$ sketched in figure 10.1 allows for oscillatory solutions at large radii. The outer boundary condition to be imposed naturally depends on the physical problem that we are considering.

In using our model equation in the context of the dynamics of galaxy disks, we can imagine the situation where no energy sources are available from the outside. Obviously, if we consider a galaxy that is not isolated, this point of view might be changed. Therefore, much as in the case of the processes of α-decay in atomic nuclei, in the discussion of certain effects created by an obstacle to water waves, or in the study of some drift waves in plasma physics, we come to the conclusion that the appropriate boundary condition at large radii is that of an *outgoing* wave, that is, the radiation boundary condition. This condition states that, out of the two independent oscillatory solutions available at large radii, we should pick only the one that carries energy outward in order to prevent the system from absorbing energy from the outside.

For the specific problem of spiral structure in galaxies, an additional physical argument recommends the use of such a radiation boundary condition, and this is, strictly speaking, outside the domain of the simple fluid model that the equations describe. The point is that any outgoing wave signal in the galaxy disk is expected to be eventually absorbed, either at the outer Lindblad resonance in the collisionless stellar system, or by turbulent dissipation in the gas layer. Thus no signals from the outside are expected (which might occur if the galaxy had a finite edge). This boundary condition can be implemented only by *trailing* waves in a galaxy disk, and this gives a natural and physical explanation for the prevalence of trailing structures in spiral galaxies.

10.1.3 Overreflection

The above-described radiation boundary condition opens the way to an important amplification mechanism known as "overreflection." If we consider the double turning point at r_{co} as representative of the situation in a galaxy disk in the vicinity of the corotation circle, we should keep in mind that the region at $r < r_{co}$ is a region of negative density of wave action, while the region outside has opposite behavior. Then, from the propagation point of view, we can visualize a signal coming from the center of the galaxy and being partly reflected back and partly transmitted across r_{co} as an outgoing wave (as required by the radiation boundary condition). Naively, we might expect both the outgoing wave and the reflected wave to be weaker than the signal originally launched toward the corotation circle. However, because of the negative sign of the action density associated with waves at $r < r_{co}$, the direction of energy and angular momentum fluxes for $r < r_{co}$ is *opposite* to the direction of group propagation. Therefore, from the action (or angular momentum) flux point of view, in order to conserve the action flux across r_{co} the reflected signal must be *stronger* than the original wave impinging on r_{co}! This amplification process is depicted in figure 10.2 (see also figure 4.1), where the arrows denote group propagation. Clearly, the overreflection is more efficient if propagation across corotation is encouraged (i.e., in the case where the g-curve is lowered at r_{co}, as in the case shown in the middle frame of figure 10.3), or is

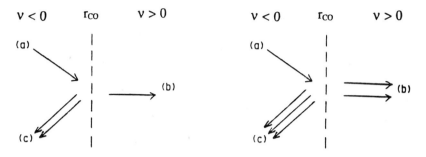

Figure 10.2
Process of overreflection illustrated for two cases of different levels of local stability at corotation. *Left:* Case of local marginal stability (see figure 10.1 or left frames of figures 9.1 and 10.3). *Right:* Case of a system locally unstable at corotation (see middle frames of figures 9.1 and 10.3); for a description, see also the caption to figure 4.1).

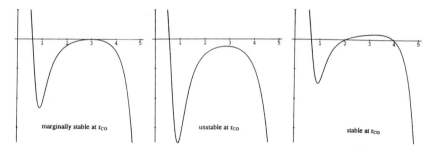

Figure 10.3
Function $(-g)$ can be seen like the potential for a particle with zero energy. The plots given here actually reproduce the function indicated in eq. (10.2) (regime A) for the same models used to draw the propagation diagrams in figure 9.1, following the same sequence from left to right; there is a small difference, in that here the departures from unity of Q_{OD} have been slightly exaggerated, in order to better appreciate the changes induced by modifying the conditions of local stability at corotation ($r_{co} = 3$, as in figure 10.1). This figure can also be used to describe possible cases related to regime B (section 10.2.2).

still present, but weak, if the waves have to make their way across r_{co} through tunneling (i.e., in the case where the g-curve is raised as in the case shown in the right frame of figure 10.3). Thus the *amount* of overreflection is not determined by conservation of wave action alone. Amplification takes place because energy is transferred from a negative energy density region ($r < r_{co}$) to a positive energy density region ($r > r_{co}$), as required by the radiation boundary condition.

It is clear that the free energy tapped in such a process is associated with the presence of a shear flow, that is, with the differential rotation.

In a sense, this collective behavior tends to accomplish what viscosity would do in an ordinary fluid, that is, to redistribute angular momentum in the system toward a state of rigid, solid body rotation. Thus it is no surprise that some analogies could be drawn at this point with processes occurring in shear flows in the context of hydrodynamics and meteorology, or taking place in sheared magnetic configurations in the context of plasma physics.

10.1.4 Self-Excited Discrete Global Spiral Modes

At this point we have all the ingredients necessary to justify the existence of a discrete spectrum of unstable (i.e., self-excited) global spiral modes of the trailing type. Indeed, we expect that only a few special choices of ω will shape the g-profile in such a manner for the solution to satisfy both the regularity condition at $r = 0$ and the radiation boundary condition. In order for this to occur, based on experience from other research areas in physics, we anticipate a condition on the integral of $\sqrt{g(r, \omega)}$ between the two turning points to be an appropriate multiple of $\pi/2$. On the other hand, in contrast with other, more common physical examples, the eigenvalues are bound to have an imaginary part corresponding to *growth in time* in order to fit in with the amplitude requirement imposed by overreflection. In more intuitive terms, as originally described by Mark [12, 13], the system behaves like a cavity for a laser process, whereby a wave-signal gains in amplitude as a result of overreflection at each cycle at a rate determined by the bounce time associated with the feedback at r_{ce}. Thus the growth rate of the mode, for a given shape of the g-profile, is inversely proportional to the group propagation time required by a wave signal to move from r_{co} to r_{ce} and back to r_{co}.

This picture is expected to hold to the extent that no impediment arises to limit the feedback process from the central regions of the disk. We shall see in section 10.4 that inner Lindblad resonance can, in many cases, provide such an impediment, and this has the welcome role of limiting the number of unstable spiral modes in a galaxy disk.

10.2 The Two–Turning Point Problem for A-Modes and B-Modes

10.2.1 Low-J Limit

We have seen that in the low-J limit the basic dispersion relation is eq. (9.7), which leads to propagation diagrams of the type shown in figure 9.1. Thus we expect that unstable global spiral modes can occur based on trailing waves only. Short trailing waves can carry angular momentum to large radii, while long trailing waves are available to justify feedback at r_{ce}. The situation is already self-evident in the structure of the propagation diagram.

In order to be more quantitative in our description, we can record here the second-order differential equation that can be derived in this regime, which is to lowest order in the asymptotic analysis of tightly wound waves [8]:

$$\frac{d^2u}{dr^2} + \frac{1}{r^2\epsilon^2(r)} \left(\frac{1}{Q^2} - 1 + \nu^2 \right) u = 0, \tag{10.2}$$

where ϵ, Q, and ν have been defined in chapter 9, and the function u is related to the unperturbed enthalpy h_1 by a transformation that "subtracts out" the basic wave number $|\hat{k}_0| = 2/Q^2$ (see eqs. (9.11) and (9.12)). Thus the turning points of eq. (10.2) (i.e., the zeros of the g-profile specified in this case) correspond to the vanishing of the square root in eqs. (9.11) and (9.12), that is, to the meeting of short and long (trailing) waves.

As in other WKB problems the differential form (10.2) contains all the information present in the associated algebraic dispersion relation. However, the differential form is *regular* at the various turning points and allows us to calculate the solutions in their vicinity, that is, to calculate properly the results of linear wave-wave interaction.

Feedback at r_{ce} occurs because an incoming short wave is refracted back to the outer disk, as a long wave, by the "Q-barrier" that was argued on physical grounds in chapter 7. Such a long wave reaches corotation and is overreflected into a pair of short waves; the one propagating inward, in the region inside r_{co}, repeats the process that maintains the mode, the other propagating outward, at $r > r_{co}$, carries angular momentum away as required by the radiation boundary condition.

The faucet that determines the amount of overreflection (i.e., the amount of energy tapped by the mode) is clearly regulated by the Q-parameter value at $r \approx r_{co}$. For $Q > 1$, leakage of angular momentum is small, via tunneling, and the overreflection is very mild (see right frames of figures 9.1 and 10.3). It can be shown that for $Q = 1$ the overreflected short trailing wave is twice as strong (in the square of the amplitude) as the incoming long wave (see left frames of figures 9.1, 10.2, and 10.3). It should therefore be emphasized that a disk *"locally"* stable everywhere with respect to (axisymmetric) Jeans instability ($Q \geq 1$) can

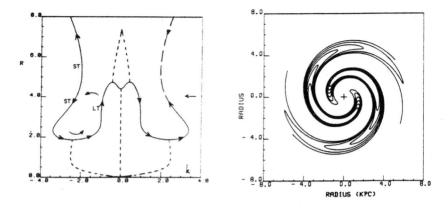

Figure 10.4
Propagation diagram (left), based on the cubic dispersion relation, showing the possibility for an all-trailing wave cycle approximated reasonably closely by an equation of the type derived for regime A. Corotation is at 4 kpc, marked by a horizontal arrow; the corresponding spiral mode (A-type) is shown on the right. [1]

be *globally unstable* (because of overreflection) to the onset of large-scale spiral modes.

The resulting modes can be called "A-modes" in that their morphology is expected to be all-trailing with no bar feature in the middle. The discussion given above should serve as a reasonable approximation even when J is not so small, provided we are in the regime where $J < 1/2$, that is, in region A of the J-Q diagram (see figure 9.3). Of course, finite J modifications are expected, as indicated in the propagation diagram of figure 10.4 based on the more general cubic dispersion relation. Still we expect that the essentials of the mode excitation and maintenance can be captured by the short-long wave cycle, as approximated by eq. (10.2).

10.2.2   Open Wave Limit

In the open wave limit of $J = O(1)$, the relevant wave cycle must include waves of both the leading and trailing type. In such a regime an ordinary differential equation can be derived to describe the process [1, 5]:

$$\frac{d^2w}{dr^2} + g_{\text{open}}(r, \omega)w = 0, \tag{10.3}$$

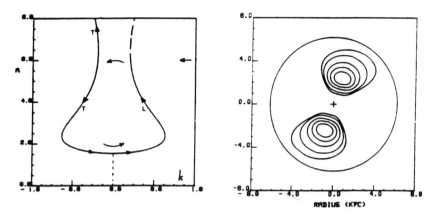

Figure 10.5
Propagation diagram (left) based on the cubic dispersion relation, showing the leading-trailing wave cycle approximated reasonably closely by an equation of the type derived for regime B. Corotation is at 6 kpc, marked by a horizontal arrow; the corresponding bar mode (B-type) is shown on the right. [1]

where now w represents the full perturbation (no basic wave number is subtracted out). The two oscillatory solutions allowed by eq. (10.3) are the leading and the trailing wave and correspond to the only real root available for the cubic dispersion relation in this regime of high J.

This setup for mode maintenance and excitation is well represented by the propagation diagrams found in this regime (see figure 10.5). A trailing wave moving toward the center of the galaxy is refracted back as a leading wave signal. Such an outgoing wave reaches corotation and is there overreflected (the term often used for this process of conversion of leading into trailing waves is *swing-amplified* [17]) into a pair of trailing waves. The signal that goes back toward the center repeats the process and maintains the mode, while the outgoing trailing signal carries angular momentum away as required by the radiation boundary condition.

The overall shape of the mode is biased, especially outside corotation, to the trailing side. However, due to the superposition of low-k waves of both leading and trailing type, the mode morphology inside r_{co} is expected to display a two-lump structure (for $m = 2$), that is, to be of the general bar type. Thus these modes are called "B-modes." This general description is expected to apply in the regime of high-J in region B of moderate instability of the (J, Q) diagram (see figure 9.3). Indeed, for these modes the faucet is expected to be regulated not only by Q, as in the low-J regime, but also by J, exactly as described by region B in the (J, Q) diagram.

10.2.3   Quantum Condition and Discrete Spectrum of Self-Excited Modes

It would be instructive to carry out in detail the calculation of the discrete spectrum of unstable modes, and readers are referred for simplicity to the specific case of the low-J limit in the case where $Q_{OD} = 1$ [8]. Here we may summarize the main results of the analysis by saying that the second-order ordinary differential equation can be solved separately, in two different intervals, across each of the two turning points, in a way that implements the above-described boundary conditions. This produces two apparently different expressions for the solution in the interval $r_{ce} < r < r_{co}$. Within a proportionality factor the two expressions for the solution should be the same because they are supposed to describe the same eigenfunction. Such identification, or *matching*, yields the so-called quantum condition:

$$e^{i\pi} = \sqrt{2}\exp\left\{2i\int_{r_{ce}}^{r_{co}}\sqrt{g}\,dr\right\}. \tag{10.4}$$

This solution with the factor $\sqrt{2}$ quantifies the effects of overreflection and is strictly applicable to the case of local marginal stability at $r \approx r_{co}$. For cases different from marginal stability (see the middle and right frames of figure 10.3) in the corotation region, the double zero becomes a pair of close-by zeros yielding a similar relation with a different "amplification" factor.

The global dispersion relation, eq. (10.4), requires the existence of complex eigenvalues. Recalling that the argument of the integral is indeed related to the wave number and expanding to first order in a Taylor series with respect to the imaginary part of $\omega = \omega_R + i\omega_I$:

$$k \simeq k(\omega_R, r) + i\left(\frac{\partial k}{\partial \omega}\right)_{\omega_R}\omega_I, \tag{10.5}$$

we can reduce eq. (10.4), for $|\omega_I| \ll |\omega_R|$, to

$$\oint k(r, \omega_R)\,dr = (2n + 1)\pi \tag{10.6}$$

and

$$\gamma \oint \frac{dr}{c_g} = \frac{1}{2} \ln 2, \tag{10.7}$$

where the loop integrals are taken as a full trip between r_{ce} and r_{co}, and n is a nonnegative integer. The growth rate is given by $\gamma = -\omega_I$.

The first relation determines the real part of the eigenvalues ω (i.e., the allowed values for the corotation radius) and is a Bohr-Sommerfeld type of condition. Note that the loop integral gives the area of the region of the propagation diagram "enclosed" by the relevant wave cycle (cf. figure 9.1). From this setup it is apparent that the first mode ($n = 0$) is the fastest and has the smallest corotation circle. Higher n modes are slower rotating, with corotation circle further out.

The second relation involves, as anticipated, the group velocity propagation time in a full cycle between the two turning points. The growth rate would be faster if the $\sqrt{2}$ factor were modified, for models that are locally unstable at r_{co}.

A solution of this type can be justified in each of the two simple regimes A and B. To be sure, we should keep in mind that both the derivation of the "quantum condition" and the reduction to the second-order equations in the two regimes of A-modes and B-modes involve a number of simplifications with respect to the full mathematical problem. Thus we would like to use these conclusions mostly as interesting, simple guidance for a more thorough and exact numerical treatment and especially as a tool to understand the basic processes by which modes are excited and maintained in galaxy disks. A simple inspection of the propagation diagrams defined by the more general cubic dispersion relation immediately reveals that actual modes can benefit from a variety of wave channels and processes that goes well beyond the two idealized regimes identified so far.

10.3   Exact Modes from Numerical Integration

10.3.1   An Integro-Differential Eigenvalue Problem

If we consider the set of equations describing a one-component, zero-thickness, inviscid fluid characterized by a barotropic equation of state in the presence of an immobile spherical bulge-halo mass distribution, then the linearized equations for elementary density waves (which leave the disk geometry unperturbed) with phase $\exp[i(\omega t - m\theta)]$ can be reduced to an integro-differential equation with respect to the radial variable r. An asymptotic analysis of this "exact" eigenvalue problem leads to the approximate dispersion relation discussed in chapter 9. As stressed several times on previous occasions, a key nontrivial step at the basis of the derivation of the dispersion relation is the reduction of the long-range gravity law to a local relation between the perturbed potential Φ_1 and the perturbed density σ_1.

As is generally the case with asymptotic analyses, the best and indisputable way of proving the merits and of defining the range of validity of the asymptotic results is to compare the asymptotic theory with the results of an "exact" numerical integration of the basic equations. This indeed has been performed on the basis of the integration of the full set of equations for the fluid model for thousands of galaxy models, and the results have been very satisfactory. We should mention here that such a numerical investigation grew out of a code initially developed by R. Pannatoni [14] (the properties of the mathematical problem and the structure of the code are especially well described in [15]). This forms the basis for studies by Thurstans [16] and Lowe [9]. Different numerical procedures have been explored by Haass [7].

For a given value of m, the numerical code identifies special complex values of ω (eigenvalues) which admit nontrivial density perturbations (eigenfunctions), that is, it identifies global modes. Each solution corresponds to a spiral or barlike density perturbation rotating rigidly at a "pattern" speed $\Omega_p = \omega_R/m$ and growing in time exponentially as $\exp(-\omega_I t)$. These are the counterparts, within an "exact" linearized fluid theory, of the global modes described earlier in this chapter.

10.3.2 Modal Survey

An extensive modal survey has been carried out by numerical integration of the exact equations for linear perturbations on the family of one-component, zero-thickness fluid galaxy models described in chapter 7 [4, 5]. Most of the astrophysical implications of such a survey have been presented in part I of this monograph, especially in chapter 4. There it has been emphasized how, following the properties of the basic models in the available parameter space under the constraint of *moderate instability* (i.e., essentially along the dotted region from A to B in the (J, Q) diagram), the survey has indeed identified essentially all the key morphological types observed in the Hubble classification scheme. For simplicity, a few mode prototypes have been singled out, as shown in figure 10.6. Equilibrium states subject to too violent instabilities are unlikely to correspond to *present* conditions of observed spiral galaxies; more likely, they describe fast-evolving situations that may have occurred in the past.

Note that the various unstable modes that are found in the survey are characterized not only by their specific morphology, as shown by the perturbed positive density contours, but also by their corotation circle (usually displayed as a dotted circle; see again figure 10.6). Thus the modes that are excited have a well-defined scale with respect to the scale length of the underlying basic state. It is at this stage that the results of the survey, combined with the dynamical considerations involved in the modeling process (see chapter 7), show that bar modes tend to be characterized by $r_{co} \approx 1\text{--}2h_*$ (with h_* the exponential scale of the underlying stellar disk), while normal spiral modes, which owe much of their excitation and support to the extended gas component, are expected to have $r_{co} \approx 3h_*$.

The modes found in the survey have been "diagnosed" in a variety of ways. On the one hand, for each mode, the propagation diagram has been drawn based on the cubic dispersion relation described in chapter 9 (see figure 9.2). This has been found to be a simple, excellent tool for comparison of the exact eigenmodes with the expectations of the asymptotic theory as outlined in sections 10.1 and 10.2. Another interesting diagnostic tool, which does not depend on a specific asymptotic theory, is that of the so-called α-spectra. For a given m-component in an azimuthal Fourier analysis of the spatial structure of the perturbed density $\sigma_1(r, \theta) = \tilde{\sigma}_1(r)e^{-im\theta}$, we can introduce the quantity

$$\sigma(\alpha) = \frac{1}{2\pi} \int_0^\infty \tilde{\sigma}_1(r)e^{i\alpha \ln r} \frac{dr}{r}, \tag{10.8}$$

which corresponds to an expansion in logarithmic spirals. We recall that logarithmic spirals are characterized by a constant pitch angle given by the relation $\tan i = m/\alpha$ (see chapter 8). Thus the quantity α/m is related to the quantity rk/m used in chapter 9 to distinguish between the regime of open waves from that of tightly wound waves. Recall also that $\alpha < 0$ corresponds to the trailing winding direction. Thus the quantity $P(\alpha) = |\sigma(\alpha)|^2$, which we call "$\alpha$-spectrum," gives an interesting measure of the "wave number" content of a given disturbance. The α-spectrum is an interesting diagnostic tool that can be applied both to calculated global modes and to observed spiral structure (as in the studies illustrated by figure 8.2); it does not require that the amplitude of the perturbation $\tilde{\sigma}_1(r)$ be small.

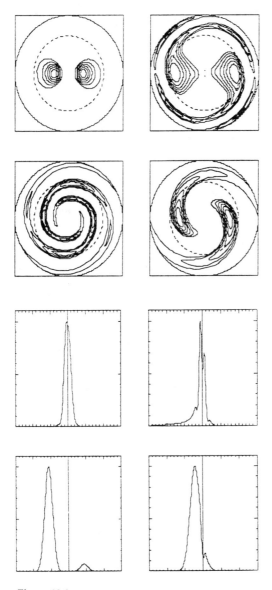

Figure 10.6
Mode prototypes from a numerical survey of global modes in fluid models of galaxy disks [5]. The four frames on the top repeat the perturbed density contours shown in figure 4.5; the related propagation diagrams were shown in figure 9.2. The four frames on the bottom give the α-spectrum for each mode prototype; on the horizontal axis α/m runs from -15 to $+15$.

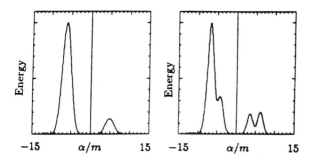

Figure 10.7
Alpha-spectra for two A-type modes in the same galaxy model of the numerical survey [5]. The spectra show contamination by leading waves; note, however, the clear appearance of the long wave branch in the spectrum of the second mode, shown on the right.

The combination of these diagnostic tools gives good support to the picture of mode excitation and maintenance outlined in sections 10.1 and 10.2 and exemplified by the two limiting cases of A- and B-modes. In the regime of A-modes the normal spiral structure obtained is closely approximated by that derived from the analysis of section 10.2.1, and the resulting pattern is closely approximated by the short wave branch of the quadratic dispersion relation (as argued originally in the first applications of the density wave theory—see chapter 5). These normal spiral modes gain support from the short-long wave cycle. One might naively expect that short and long waves would appear as separate peaks in $P(\alpha)$ for negative values of α, but more often we see only a single peak. The explanation for this apparent paradox is that the two contributions may merge and need not appear as separate features, especially because the spectrum is biased toward the short wave branch (on which overreflection operates). The α-spectrum also shows that there is always some contribution from *leading* waves (see figure 10.6), which is not unexpected, as a "coupling of resonant cavities." In some cases, especially for the higher modes, the role of the long wave branch is explicitly brought out by the α-spectrum (see figure 10.7).

For the regime of open modes (the B-modes), one may worry that, even if the propagation diagrams based on the cubic dispersion relation appear to support the picture of excitation and maintenance for the modes outlined in section 10.2.2, the asymptotic analysis may be insufficient or of little use for quantitative purposes. This is a legitimate worry because the analysis is somewhat stretched and the very

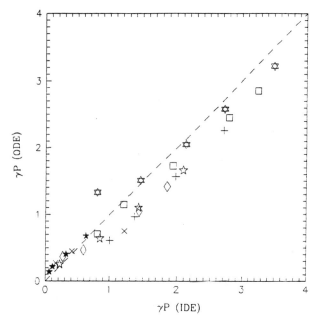

Figure 10.8
Test of the asymptotic theory described in sections 10.1 and 10.2 [5]. For a number of models in the numerical survey, the growth rate obtained from the ordinary differential equation (ODE) of the open wave limit is compared with the exact results obtained from the integro-differential equation (IDE). Differential symbols identify different values of Q_{OD} of the models for which the test has been made, with Q_{OD} ranging from 1 (open squares) to 1.4 (filled stars).

open waves most likely defy any local treatment, with the gravity as a true long-range force. A posteriori, we find that the overall agreement with the asymptotic analysis is actually very satisfactory, probably better than originally hoped for. This point is well illustrated in figure 10.8, where the growth rates for B-modes as calculated from the simplified ordinary differential equation (10.3) are compared with those computed from the full integro-differential equation, for several models with different location in the (J, Q) diagram.

Finally, we should stress that the survey of modes shows that both for A-modes and for B-modes one easily finds amplitude modulations along the spiral arms; such modulations are naturally interpreted as the result of interference patterns between the various waves that support the global modes. This morphological aspect has been recognized to be an interesting point of comparison with observations of spiral galaxies, which indeed often display such regular modulations along the arms (see chapter 5).

10.4 Inner Lindblad Resonance Limits the Number of Unstable Modes

The analysis of mode excitation and maintenance as given in sections 10.1 and 10.2 does not include the role of resonances and other stellar dynamical effects properly; for this purpose, it has to be supplemented by a few physical considerations. Resonances appear as singularities in the integro-differential system of equations. The corotation resonance ($\nu = 0$) has been found to have only a modest impact on the problem of large-scale spiral modes for a reasonable choice of the basic state. Other stellar dynamical effects, such as the role of pressure anisotropy, which is characteristic of the stellar disk (while the pressure in our fluid model is isotropic), can also be considered and taken into account. Here we focus on the role of the inner Lindblad resonance because it has a major impact on the application of these studies.

Detailed dynamical studies of the Lindblad resonances show that a *stellar* disk acts as an essentially *perfect wave absorber* at such resonance locations [10, 11], while a *fluid* disk behaves as an *imperfect* absorber only. The collisionless stellar disk is, in other words, more efficient in the process of converting wave energy into random star epicyclic motions. The stars involved in such resonant absorption are those contained in an annulus about two epicycles wide around the resonance radius. These properties have a major impact on the expected set of unstable spiral modes in a galaxy disk. Note that if inner Lindblad resonance occurs, the ILR region of the disk is expected to be star-dominated. Dynamical studies also show that the fluid model gives generally a good representation of a stellar disk in the wave propagation region away from resonances.

If, for a given mode found from the asymptotic analysis of section 10.2 or from the numerical survey of section 10.3, the location of ILR turns out to be at $r \gtrsim r_{ce}$ (i.e., if ILR is "exposed"), the mode should be discarded as a damped mode. Indeed, such an occurrence would shut off the feedback required for the mode to be maintained (see section 10.1.1), thus destroying the wave cycle that determines global unstable modes. It is very easy to show (see chapter 7) that for reasonable galaxy models, all the modes with $m \gtrsim 3$ should be damped by this mechanism. Indeed, from the character of the "shear diagram" (see the example of figure 2.9) we expect all $m = 1$ modes to be ILR-free, $m = 2$ modes to be generally ILR-free, and $m = 3$ modes to be generally damped by ILR effects. The same mechanism operates in selecting out the higher n-modes (see eq. [10.6]), because of their lower pattern speed (which makes it easier for the intersection of the curve $\Omega - \kappa/m$ with the horizontal line at Ω_p to occur at $r > r_{ce}$). Thus, even if the integration of the fluid model may formally yield several unstable modes (because the fluid disk is not so much under the influence of ILR), the actual number of modes to be retained as applicable to real galaxy disks is much smaller, often limited to $m = 1$ and $m = 2$.

It is clear that the previous argument holds to the extent that the large-scale spiral structure that we are modeling is mostly star supported (at least for $r < r_{co}$). Thus we expect that disks that are gas-rich (especially some Sc galaxies) would have higher-m, large-scale spiral structure showing up in a prominent manner. In some of these cases, this fluid excitation would be able to drag a considerable fraction of the cool stellar disk into such multiple-armed spiral structure.

From the observational point of view, as noted in chapter 5, there is striking support to the above-described picture. Although images in the optical wave bands dominated by the Population I disk, especially in the B-band, often show large-scale, high-m structures in gas-rich galaxies (see M101, NGC 628, NGC 309), images in the infrared that probe the underlying stellar disk are generally dominated by $m = 2$ or $m = 1$.

The occurrence of high-m, large-scale spiral structure in n-body simulations leaves little doubt that, in many respects, n-body simulations display a fluidlike behavior as a result of their limited resolution in phase space.

Finally, we should note that, when the Lindblad resonance occurs too close to the corotation circle (as may happen for $m \gtrsim 5$, say), so that a single star can be orbiting close both to r_{co} and to r_{OLR} (or r_{ILR}), then incoherent, stochastic behavior is expected in the stellar disk, which gives one more reason to argue that large-scale, high-m spiral modes should not be excited.

10.5 Nonlinear Evolution

The domain of nonlinear evolution is a broad research area that is only at its beginning. We briefly outline here below three aspects of the problem that have been given attention from the physical point of view and, to some extent, from the point of view of a quantitative analysis.

10.5.1 Saturation of Moderately Growing Modes

The exponential growth found in the linear modal analysis is expected to be "saturated" by nonlinear effects so that the resulting global structure equilibrates at finite amplitudes. A natural way for the process to occur is via shock dissipation in the cold interstellar medium (see chapter 3), for which we may set up a model equation of the Landau type:

$$\frac{dA}{dt} = \gamma A - S, \tag{10.9}$$

where A represents the amplitude of spiral structure, γ its growth rate, and S (a nonlinear function of A) the damping by the shocks.

Another independent way for the modes to "saturate" could be by losing orbital support at finite amplitudes (compare the discussion in section 8.6). In other words, beyond a certain amplitude the stars may be unable to support the spiral structure, thus providing some kind of saturation mechanism for the growing modes even in the absence of shocks in the gas component. Some smooth-arm spiral galaxies with a grand design appear to be gas-poor.

10.5.2 Self-Regulation of the Basic State for Normal Spirals

In chapter 3 we presented a physical discussion of the process of self-regulation by which the cold interstellar medium is thought to guarantee long-lasting conditions for the excitation of large-scale normal spiral modes. This means, for a two-component system of stars and gas, the possibility that rapid cooling in the dissipative gas compensates for the secular heating of the stellar disk in such a way that the disk is maintained, in the outer parts, close to the margin of axisymmetric instability. Figure 10.9 illustrates the process as computed on the basis of a simplified set of nonlinear evolution equations for the parameters of the basic state:

$$\frac{d \ln Q_*}{dt} = f_* > 0 \tag{10.10}$$

$$\frac{d \ln c_g}{dt} = -p + f_g \tag{10.11}$$

together with the relation that identifies the effective parameter Q for a two-component disk in terms of the properties of the two components [6]. Here $Q_* = c_* \kappa / \pi G \sigma_*$ represents the value of the axisymmetric stability parameter referred to the stellar component alone, p represents

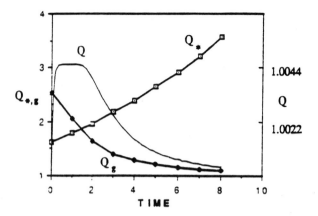

Figure 10.9
Demonstration of the mechanism of self-regulation [3]. A simple set of nonlinear evolution equations shows how a disk can be self-regulated with an effective Q very close to
unity, even when the Q-parameter for each of the two separate components (stars and
gas) is significantly evolving in time and different from unity. By the time the stellar disk
has heated up to the levels shown at the end of the integration, the stellar disk is too hot
and has become decoupled from the gaseous disk, and the galaxy is expected to have
become flocculent.

a fast, cooling term for the dissipative gas, c_g is the equivalent acoustic speed of the gas, and f_*, f_g are two heating functions that are very
large for $Q < 1$ and turned off for $Q \gtrsim 1$. The figure illustrates how the
system can self-regulate at $Q \approx 1$ even when Q_* and Q_g have, separately, sizable evolution. A more realistic physical description of self-
regulation should incorporate several other effects, such as the role of
star formation, supernova explosions, and a more detailed picture of
the various cooling and relaxation processes that may be considered.

10.5.3   Angular Momentum Transport by Spiral Torques

A third aspect of nonlinear evolution considers the effects of the large-
scale spiral torques on the properties of the basic state. The redistribution of angular momentum might lead to rapid changes in the disk
and even modify the properties of the rotation curve. A preliminary estimate for the timescale of this nonlinear evolution for normal spiral
modes suggests that the effects involved, using parameters appropriate for currently observed spiral structure, should be very slow and
significant only beyond the Hubble time [2]. To be sure, faster evolution is expected for open structures, and therefore for barred spirals.

However, the very rapid evolution suggested by some n-body simulations is probably due to the small number of particles used and to the inadequacy of the models being considered.

10.6 Basic Requirements for the Modal Approach to be Dynamically Viable

When modal and nonmodal studies are compared (see chapter 11), it is important to keep in mind that the modal description is applicable only if some basic requirements are met. These requirements, summarized below, are naturally satisfied in galaxy disks but often do not hold in some dynamical models discussed in the literature.

The system that we have in mind for the description of large-scale spiral structure is characterized by a discrete spectrum of moderately growing modes. We have noted that the outer disk is a source of excitation for normal (unbarred) spiral modes because the physical conditions there can be kept sufficiently cool by the abundance of cold interstellar gas (self-regulation; see sections 3.4 and 10.5.2). Thus one may imagine the outer disk as a source of wave signals propagating inward. It is essential for the global modes to be properly formed and maintained that the central regions provide a feedback (see section 10.1.1), so that wave signals are returned toward the outer disk. The global modes are those structures for which the returned waves arrive with the proper phase in the outer disk, so as to collectively transfer angular momentum outward ("radiation boundary condition"; see section 10.1.2). This picture is physically justified only if the group propagation time is *short enough* (with respect to the Hubble time); otherwise, the galaxy may not have had sufficient time for the modal behavior to be established (see section 7.2.6). In this respect, the spiral structure in some very large galaxies, such as UGC 2885, for which the rotation period at the periphery may be of the order of one billion years, may not have had time to fully settle into modal behavior (although the symmetry of the observed structure in UGC 2885 would indicate that, even under such marginal conditions, global modes are actually standing out; see chapter 5).

In the monograph we have also emphasized the important beneficial roles of the gas as a cold dissipative component for the establishment of normal (unbarred) spiral structure. In particular, gas helps to equilibrate the growing modes into nonlinear structures at finite amplitudes (see chapters 3 and 4). To be sure, because only some of these important nonlinear aspects of modal evolution have been worked out quantitatively, this is one research area where further investigations are desired (see section 10.5).

The modal description may be inconvenient for systems satisfying the requirements stated so far if too many self-excited modes are present. The presence of too many modes could hardly be reconciled with observed regular spiral structure. If this were the case, we might have to invoke external help to organize spiral structure into a coherent bisymmetric structure of the kind often observed. For realistic galaxy disks, absorption at the inner Lindblad resonance (which prevents feedback) is probably the key mechanism that limits the number of self-excited global modes (see section 10.4). This point of view is supported by the frequent observation in the optical of multiple-armed and less regular spiral structure (in the gas-dominated Population I layer, ILR is partially transparent to density waves) and by the more regular and low-m spiral structure observed in the infrared (total absorption of density waves occurs at ILR in the star-dominated Population II layer). Thus we would expect that a gas-rich galaxy would be very regular in the optical only if some extra coherence is supplied by external mechanisms (this is probably the case of M51).

Another important factor to keep in mind, which contributes to limiting the number of modes present, is the specific gas distribution, relative to the stellar distribution, that occurs in a given galaxy disk. The conditions for proper self-regulation and excitation of unbarred spiral modes are met only in the outer disk. Higher modes (high n modes; see section 10.2.3), besides running into the possible problem of ILR (because they are slower rotating), would have corotation so far out that, even if gas were present, the stellar disk would be essentially absent; thus no support would be available for the maintenance of large-scale structures of these higher-n modes. In conclusion, the establishment of large-scale modes requires that several factors cooperate properly, so that very few large-scale modes can be practically established (see the modal survey described in chapter 4 and section 10.3). Finally, nonlinear mechanisms may help selecting, out of the very few self-excited modes that are found in the linear theory, the mode or the couple of modes that eventually dominate a very regular observed spiral pattern. This is another area where further quantitative investigations are obviously needed.

Under these circumstances, we may then approximate the current observed structure with the help of linear modes by arguing that large-scale structure is probably only gradually evolving when it is very regular (see section 4.4) and that the dynamical system has settled down into a state in which a single mode or a couple of modes might be dominating. The evolutionary process of settling down from some initial conditions is bound to be complicated, and we have no intention of pursuing it in the present monograph. Rather, we turn to the observational data for support that the modal description is indeed applicable to a *statistical majority* of galaxies. Indeed, our conclusions provide a simple theoretical framework for the interpretation of the various morphological categories of spiral galaxies (see chapter 4, especially section 4.7, and the recent infrared studies described in chapter 5).

10.7 Concluding Remarks

Ever since the earliest attempts to develop the density wave theory into a quantitative form, a semiempirical approach has been adopted. We first checked the quantitative nature and the general magnitude of the effects of the spiral gravitational field to make sure that it introduces only a small, though finite, perturbation over the general circular motions in the galactic disk. We then checked some of the observational data in the Milky Way and in nearby galaxies. Thirty years later, we have available a great deal of observational data to be compared with a continually improving theory that is still constructed on the same basic perceptions. In particular, the modal approach has been found to be compatible with a wide range of observational data, it has provided an effective general framework for the understanding of the morphological and luminosity classification of galaxies, and it has also provided proper insight and explanations for the specific data of several individual galaxies. At this time, the welcome addition of reliable infrared data is giving further empirical support to the theory (see chapter 5). The regularity of the global patterns found gives support to the modal approach described here; that is, we are dealing, in most cases, with long-lasting and slowly evolving global structures that are primarily standing wave patterns maintained by intrinsic mechanisms.

The origin of such density wave patterns would generally be traced back to the evolution of the galactic disk since primordial times. Although some of the observed global spiral structures might be thought as initiated from stable featureless disks through recent encounters with neighboring galaxies, most such encounters probably involve galactic disks with preexisting spiral structures, especially if the galaxy in question is gas-rich. One might expect a fast-evolving global structure if it were generated, under appropriate circumstances, by a recent encounter, but such scenarios would not be expected to be observed frequently, and even such a spiral structure would eventually evolve toward a slowly changing standing wave pattern determined by intrinsic characteristics of the galaxy.

It is the purpose of the following chapter to discuss and to clarify some of the subtle issues related to the evolutionary process and to briefly describe some alternative scenarios that should be explored in further quantitative detail.

10.8   References

1. Bertin, G. 1983a, in **Internal Kinematics and Dynamics of Galaxies**, IAU Symp. 100, ed. E. Athanassoula, Reidel, Dordrecht, p. 119.

2. Bertin, G. 1983b, *Astron. Astrophys.*, **127**, 145.

3. Bertin, G. 1991, in **Dynamics of Galaxies and their Molecular Cloud Distributions**, IAU Symp. 146, ed. F. Combes and F. Casoli, Kluwer, Dordrecht, p. 93.

4. Bertin, G., Lin, C.C., Lowe, S.A., and Thurstans, R.P. 1989a, *Astrophys. J.*, **338**, 78.

5. Bertin, G., Lin, C. C., Lowe, S.A., and Thurstans, R.P. 1989b, *Astrophys. J.*, **338**, 104.

6. Bertin, G., and Romeo, A.B. 1988, *Astron. Astrophys.*, **195**, 105.

7. Haass, J. 1982, Ph.D. diss., Massachusetts Institute of Technology.

8. Lau, Y.Y., Lin, C.C., and Mark, J.W.-K. 1976, *Proc. Natl. Acad. Sciences U.S.A.*, **73**, 1379.

9. Lowe, S.A. 1988, Ph.D. diss., Massachusetts Institute of Technology.

10. Lynden-Bell, D., and Kalnajs, A.J. 1972, *Mon. Not. Roy. Astron. Soc.*, **157**, 1.

11. Mark, J.W.-K. 1971, *Proc. Natl. Acad. Sciences U.S.A.*, **68**, 2095.

12. Mark, J.W.-K. 1976, *Astrophys. J.*, **205**, 363.

13. Mark, J.W.-K. 1977, *Astrophys J.*, **212**, 645.

14. Pannatoni, R.F. 1979, Ph.D. diss., Massachusetts Institute of Technology.

15. Pannatoni, R.F. 1983, *Geophys. Astrophys. Fluid Dyn.*, **24**, 165.

16. Thurstans, R.P. 1987, Ph.D. diss., Massachusetts Institute of Technology.

17. Toomre, A. 1981, in **The Structure and Evolution of Normal Galaxies**, ed. S.M. Fall and D. Lynden-Bell, Cambridge University Press, Cambridge, p. 111.

11 Comments on the Evolutionary Process

In this monograph we have followed the hypothesis that, in most galaxies, large-scale spiral structure is slowly evolving, and we have developed a modal approach for its description. In this chapter we shall briefly comment on some dynamical issues that grow out of *alternative* scenarios that might be investigated (see sections 2.4.4 and 4.8); these alternative scenarios mostly focus on *normal, unbarred* spiral structure (although one might even speculate whether barred spiral structure could be considered to be transient and driven). Most of the arguments presented in this chapter refer to normal spiral structure. One of the most appealing aspects of the modal approach is that it puts barred and unbarred spiral structure on the same footing, as long-lasting, intrinsic, self-excited structures.

11.1 Modal and Nonmodal Approaches

From a dynamical point of view, the problem of spiral structure in galaxies is generally reduced to the study of the evolution of density perturbations over an axisymmetric disk. This is a well-defined mathematical problem, although it should not be taken literally as a representation of conditions of a galaxy disk that actually occurred in the past. There is no reason to assume that galaxy disks were axisymmetric a few billion years ago and then developed spiral structure only in more recent epochs. (We will probably never be able to tell, from theoretical or from observational arguments, what such "initial conditions" were.) Rather, the axisymmetric basic state is an idealization that is performed with the goal of describing the current observed state (see section 1.4 and chapter 7) and of clarifying which configurations are favored energetically. In other words, one may argue that, no matter what the initial conditions may have been and no matter what the detailed processes of evolution may have been, galaxies have probably settled down into some "final" state (the current state), which may be seen as analogous to a state of minimum "free energy" or to an "attractor," as it is known for general nonlinear dynamical systems. This point of view is especially justified if one notes the presence of the *dissipative* interstellar medium. In addition, we recall that resonances and other collective effects can play the role of an *effective dissipation* (e.g., Landau damping at ILR) even if the stellar component, by itself, is essentially collisionless. The combination of these mechanisms severely limits the number of significant modes to a very small number, as repeatedly pointed out earlier in the monograph (especially in chapter 10). This is one of the key points in making the modal approach effective and plausible. In this respect, the analogy sometimes made between spiral structure in galaxy disks and the modes describing the transition from the Maclaurin to the Jacobi sequence of classical ellipsoids (see section 4.5) should not be interpreted literally in terms of the specific dynamical mechanisms involved in the two cases, but rather in view of the above-mentioned identification of less symmetric, energetically favored, configurations.

11.1.1   General Definitions

If we refer to the mathematical problem, or the "gedankenexperiment," of the evolution of perturbations over a given basic state, it is well known that the modal approach developed in this monograph is suitable for describing the *long-term* behavior of *general* perturbations (in terms of the properties of a few global modes); the approach is not suitable for the description of *transient* behavior and of the dynamical development as a result of some *special* or *singular* initial conditions. Alternatively, one may consider directly the *initial value problem*; that is, one may attempt to follow the evolution of the system under prescribed initial conditions. The limitations and the advantages of such modal and nonmodal approaches have been discussed in great detail in various contexts, and especially in some classical hydrodynamical studies; in particular, some of the hydrodynamical studies have focused recently on the possibility of "nonnormal" behavior (see section 11.3).

In broad terms we may state that modal and nonmodal descriptions are generally equivalent; however, it is clear that for the purpose of de-

Figure 11.1
Hurricanes present some analogies with spiral galaxies [7].

scribing some specific phenomena, one approach may be preferable to the other. For example, in the discussion of short-term weather predictions, the evolutionary, nonmodal approach is obviously the one to be used (but see figure 11.1 for an example taken from meteorology, where an element of stationarity is obviously present). Therefore, one should examine carefully whether the modal approach is indeed the one to be adopted for the problem we are investigating.

11.1.2 Nonmodal Description in the Galactic Context

We have widely discussed in the monograph the merits of the hypothesis of quasi-stationary spiral structure and of the modal approach (which are intimately related to each other). However, if one considers the possibility that large-scale spiral structure be *transient*, or if one focuses on *small-scale spiral activity*, the initial value problem, namely the nonmodal description, would be preferred. In section 11.2 we shall briefly comment on some specific scenarios that may be considered.

Here we should recall that the initial value problem, often analyzed in order to demonstrate the possibility of transient spiral activity in galaxies, is generally set up (in the existing literature) in an idealized model in the form of an *infinite homogeneous sheet* characterized by a *constant shear rate* [5, 8, 9, 10, 22]. The geometry considered is usually Cartesian, sometimes with doubly periodic boundary conditions [23]. Such a model is taken to represent a "local patch of a disk," that is, a narrow annulus or a part of it; its dynamics are generally examined in the rotating frame where the center of the patch is at rest. Transient growth and subsequent decay of "swinging wave packets" are observed in such a model, somewhat similar to the case of the process studied by Orr [18].[1]

A major open problem is the applicability of the infinite homogeneous sheet model to the *global* context of the dynamics of inhomogeneous galaxy disks. It is obvious that the spectrum of perturbations for such an ideal shearing sheet has a behavior that is completely different from that of an inhomogeneous disk with its characteristic boundary conditions. Ideally, one should match the dynamics of the *local* patch to the global context, via appropriate boundary conditions, but such matching is as yet not available (cf. appendix C of [2]). In particular, it is easy to see that short trailing wave signals in a galactic disk naturally propagate *away* from the part of the sheet that is at rest in the rotating frame (the corotation zone), and rapidly reach regions where the physical conditions are different from those assumed for the homogeneous disk.[2] In other words, the infinite shearing sheet model is inadequate for the description of *global* structures in galaxies.

Toomre's [22] example of the "fate of a swinging wave packet" (his figure 8) in the global context also has several characteristics that make it more an extension of the homogeneous sheet rather than a true realis-

---

1. Note, however, that vorticity is central in Orr's case, in contrast with the swing mechanism for galaxy disks. Here, whether the system is two-dimensional makes a great difference (see section 11.3.).

2. Note that for a wave packet the crossing time is longer than the "swing" time, but both timescales are short. The *net* overreflection factor for a swinging wave packet is similar to that for wave trains described in section 10.1.3.

tic case. Indeed, the disk model considered by Toomre has no scales (it is self-similar), the local stability parameter is taken to be constant, and ILR is "exposed" (so as to prevent feedback from the central regions); furthermore, the example illustrates the evolution of a rather arbitrarily chosen initial condition. It shows that the process of swing amplification is strongly localized in space and time, while most of Toomre's figure 8 shows the effects of group propagation across the galaxy disk. (In the same paper the effect of "tidal" driving is simulated by imposing a strong, rapidly evolving, and perfectly bisymmetric gravitational field, with conditions that are too special to be considered as representative of generic tidal encounters.)

Some of the studies demonstrating the possibility of fast evolution and transient spiral activity are based on models that are violently unstable, and, as such, these are unlikely to represent the current state of observed galaxy disks. As we have emphasized earlier in the monograph (see section 4.3), realistic basic states are likely to be subject to moderate instabilities only. Furthermore, we have shown that a large class of realistic basic states is subject to moderately growing global spiral modes.

So far there is no convincing demonstration of the existence of *realistic* basic states free from self-excited global modes and still able to display transient (driven) coherent large-scale spiral structure. A theoretical study of such a scenario would require at least a linear analysis that includes (1) physically reasonable models of the galactic disk, (2) proper boundary conditions for the perturbations in the outer galactic disk and at the center, and (3) appropriate initial conditions. Unfortunately, such a study has not yet been carried out. In somewhat different terms, one might identify a large class of models, stable from the modal point of view, and study their spectra of decaying or neutral modes. Then one might study the astrophysical viability of these models, especially in order to have physical justification of the selected basic states and in order to answer the many astrophysical questions related to spiral structure that we have formulated at the beginning of this monograph (see section 1.4).

11.2 Some Dynamical Scenarios and Questions

The study of several well-posed dynamical questions (see more examples below that summarize some points discussed earlier in sections 2.4.4 and 4.8) will lead to a better understanding of possible dynamical scenarios. Still, each scenario should be evaluated separately in order to judge its astrophysical viability and its relevance to the various important issues raised by the problem of observed spiral structure in galaxies (see chapter 1).

Many of the points raised by the following questions have been partially explored, or sometimes even prompted, by n-body simulations (see [1] and the references therein). However, as often remarked in this monograph, n-body simulations are often misinterpreted as a result of the limited number of particles used and because of the often underestimated difficulties in setting up a physically realistic model (see earlier comments, especially section 4.3).

In some dynamical scenarios, the possibility is considered that spiral structure in a given galaxy is externally excited through tidal fields (that is, differential gravitational forces) exerted by nearby objects (bound satellites or occasional encounters) and by the cumulative effects of more distant galaxies. Obviously, these external influences are present to some extent. Indeed, they would act as a ubiquitous "initiation" process for the development of large-scale structure, if one chooses to consider the mathematical problem of the evolution of the perturbations over an axisymmetric basic state (see the comments in section 11.1.1). Still, even in this case, a "final" state of quasi-stationary spiral structure might eventually set in, in the nature of a transition into the energetically favored *phase*, which is almost entirely controlled by intrinsic mechanisms and by the properties of intrinsic modes. Thus the main question is whether or by how much such tidal interactions play important roles in the long-term maintenance of the observed global structures, not just in the early phase of generation.

Various situations may be imagined. Tidal interaction may occur in the form of "tearing," with consequent excitation of rapidly evolving spiral structure (and the creation of bridges and tails), or through a "shaking" analogous to the ringing of a church bell, or as a kind of periodic "resonant" forcing. In the last case, it would usually require mediation by higher harmonics, given the low orbital frequencies involved in galaxy binaries. But, above all, one should keep in mind that these tidal interactions are most likely to be happening among galaxies with preexisting spiral structures. In such cases, there would be a modification of the spiral structure, rather than a generation of new spiral structures [17].

The cases of "tearing" and "shaking" refer to the scenario of a one-time passage of an unbounded satellite or galaxy. The case of resonant forcing refers primarily to bound systems, where continual tidal forcing is present. The selective response of the principal galaxy to these external forces is described as "resonance." With continual or persistent forcing, one would naturally expect a final state in the form of a slowly evolving standing wave pattern; that is, it is quasi-stationary over a short timescale and quasi-periodic on the long timescale. For specific observed systems (such as M51), models have been proposed based on a bound satellite (NGC 5195) inducing a repeated forcing [20]. In that case, the galaxy NGC 5194, being gas-rich, is expected to have a preexisting spiral structure.

Only rarely do we expect the possibility for genuinely transient, non-modal global spiral structure to develop from an initially featureless disk. This case will be discussed in the next section. More detailed discussions of the role of external excitation, especially in relation to the applicability of the hypothesis of quasi-stationary spiral structure, are given in [13]. Here, in figure 11.2, we recall that the possibility of both quasi-periodic and quasi-stationary spiral structure is indeed anticipated by the modal theory [14].

11.2.1 Transient, Driven Spiral Structure?

For example, we may imagine the case of a galaxy disk essentially free from intrinsic, self-excited spiral modes and consider the possibility that, with the help of external excitation (such as the one that might be provided by a suitable tidal encounter), some spiral structure is created at a given time. What would be the time evolution of such a disk? Could the structure settle down into a mode-like, long-lasting pattern in the disk? If so, how long would it take? On the basis of the previous discussions, it is clear that the proper answer would depend on the properties of *damped* (or *neutral*) *modes* of such a basic state; the development of such driven spiral structure would depend especially on the amount of dissipation present in the driven disk and on the detailed parameters that define the galactic encounter. In this monograph we have provided several arguments in favor of intrinsic excitation of spiral structure. Still, for those individual objects like M51 where a close encounter takes place, we cannot exclude the possibility that spiral structure might be tidally driven and might follow the above-mentioned dynamical scenario. Unfortunately, very little is currently known of the properties of damped modes in realistic galaxy disks.

If an encounter occurs between objects *with* preexisting intrinsic spiral structure, what would be the influence of tidal interaction on the

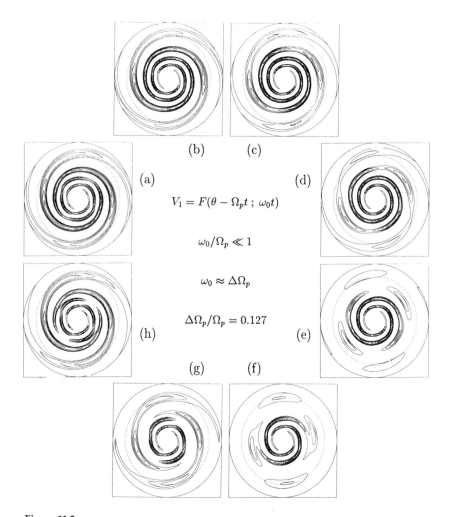

Figure 11.2
Case of quasi-periodic spiral structure, resulting from the positive and negative interference of a pair of global modes [14].

evolution of the related spiral morphologies? This is a situation that we have argued to be more relevant in the astrophysical context. Still, we have to admit that very little of this has been addressed in terms of detailed modeling and quantitative investigations. We have argued (see chapter 5) that even cases like M51 (which is gas-rich and probably subject to self-excited spiral modes) are likely to fall into this category. It would be desirable to carry out full investigations of such scenarios.

11.2.2 Recurrent, Coherent, Spiral Structure on the Large Scale?

Another specific dynamical question, often inspired by studies of the infinite homogeneous shearing sheet (see section 11.1), is whether a disk subject to transient spiral activity could produce large-scale *recurrent* coherent structure (see [15] and section 2.4.4). Here an examination of the dynamical processes involved suggests that anything that can be organized on the large scale would stand out via feedback and selective amplification as *global modes*. Otherwise it is not clear how nonmodal, rapidly evolving, *coherent* structures can be generated out of "noise." Small-scale, incoherent spiral activity is *not* expected to overwhelm the large-scale modes because these draw their strength from their coherence and large-scale organization. Indeed, empirically, large-scale structure is found to *coexist* with small-scale spiral activity.

11.3 Lessons from Hydrodynamic Stability

The difficulties of developing an adequate theory for transient growth of density wave patterns over the whole galactic disk can be appreciated by considering the situation in the context of the classical theory of hydrodynamic stability. Here the main focus is on the problem of the transition to turbulence. Although the subject is over a century old, the *linear* theory of transient growth of perturbations that satisfy boundary conditions has been adequately developed only recently (see [24]). Even here, the application of the theory to any individual category of cases still requires careful ad hoc study, although there does exist a good general framework for examining each case, together with a rigorous mathematical basis.

Transient growth of perturbations from a basic state of laminar flow is naturally well known in hydrodynamic experiments. Nonlinear processes are generally involved. In many cases, the system settles down

to a state of turbulent motion, with statistically well-defined characteristics, that includes *coherent* structures. The process of settling down to
such a state is not easily treated by analytical studies, but could be *qualitatively* described on the basis of computer simulations in some simple
cases. These simulation procedures have to be carefully examined because errors introduced by numerical approximations can have significant consequences over an extended period of time. The correctness
of the final state is best judged by comparison with laboratory experiments.

Given the wealth of experiments and of theoretical analyses in the
hydrodynamic context, one might consider proceeding by analogy arguments, applying the lessons from hydrodynamics to the galactic context. Before doing so, it is desirable to find out what hydrodynamic
stability actually tells us in order to judge the validity of such arguments. Specifically, one has to examine the *differences* between the two
contexts in order to avoid being misled by invalid comparisons.

11.3.1 Modal and Nonmodal Studies of Hydrodynamic Stability

Both the modal and nonmodal approaches in the linear theory of hydrodynamic stability were adopted a long time ago (see, e.g., [18, 19,
21]). On a global scale, the work of Rayleigh and Taylor shows that
there could be modal instability with proper boundary conditions satisfied. On the other hand, Orr found transient growth with subsequent
decay in an infinite field of flow with periodic perturbations; such
a result suggests the *possibility* of strong transient growth in a flow
bounded on two sides but does not clarify the constraining influence of
the boundaries. We should therefore refer to the recent developments
mentioned above for a definite answer.

In the bounded context (not just Orr's case of uniform shear), whenever modal instability is important, it will dominate. In some "nonnormal systems," when all the modes are decaying, there may still
be extended transient growth; but the growth is strong only if one is
dealing with three-dimensional perturbations. "When only 2D [two-
dimensional] perturbations are considered, some amplification can still
occur but it is far weaker" ([24], p. 579). The reason for this sharp contrast is well known from the physical point of view [11]. The strong
growth is provided by the physical mechanism of tilting and stretching
of vortices in the three-dimensional case; this mechanism is absent in
the two-dimensional case. Mathematically, the difference is also quite

subtle. The mathematical operator in the three-dimensional case is "exponentially far from normal" (see [24], p. 578).

The "nonnormal" systems just described are subject to "pseudomodes"; such nonnormal behavior is traced to the nonorthogonality of the set of eigenmodes that characterize the system. It is recognized that even in such nonnormal systems, a more regular modal behavior is restored through very slight changes in the basic state and/or in the boundary conditions. It is also recognized that many nonnormal systems are so close to being normal that in practice they do not show any unusual behavior. Furthermore, it remains unclear if such a nonmodal behavior can persist in the *nonlinear* context. Readers should refer to the original paper [24] for a fuller description and explanation.

Another possible source of significant transient growth in the modally stable case is the presence of a large number of important modes with a *closely packed spectrum* (cf. [16]). This is, to a certain extent, similar to the presence of a continuous spectrum of neutral modes, which is well known to be an indicator of significant transient growth.

11.3.2 Differences with the Galactic Context

Let us now return to the galactic context. There are at least the following four main dynamical differences with respect to the hydrodynamical systems known to display nonmodal behavior:

1. In contrast to the point just mentioned (last paragraph of section 11.3.1) about decaying modes, very little is actually known about such modes in realistic galactic models. From general considerations of the damping mechanisms, however, the number of significant modes in a galaxy disk is expected to be small and discrete. But a full resolution of this problem must await the type of investigation outlined at the end of section 11.1.2.

2. We are dealing only with two-dimensional perturbations because we have a different geometry for the basic state.

3. We are dealing with an *axisymmetric* system in the basic state where the modes have discrete Fourier dependence in angle. Thus the galactic problem of density waves is essentially one-dimensional. In the study of stability of flows with a similar geometry (the study of flows between rotating cylinders; see [21] and many subsequent studies, including nonlinear theory and related experiments [4, 12]), it is well known that the modal approach is entirely adequate.

4. The dominant presence of *gravity* (and *rotation*) makes galaxy disks more similar to hydrodynamical systems that are indeed characterized by modal behavior (the Taylor-Couette flow mentioned in (3) and the Rayleigh-Bénard convection in stationary fluids heated from below: for these systems, "the predictions of eigenvalue analysis match laboratory experiments" [24]).

11.4 Concluding Remarks

Transient growth is known to be important in certain dynamical contexts, in particular in the theory of hydrodynamic stability and in meteorology [3, 6]. There are many general lessons to be learned from a comparison with such investigations. The above discussions, however, show the difficulty of reasoning by analogy, if we are interested in relating those studies to the galactic context.

To sharpen our point, let us compare both the galactic context and the hydrodynamic context with the far simpler situation generally encountered in quantum mechanics (which is a *normal* linear case in the classification of [24]), where the initial value problem can be perfectly well formulated in terms of the complete set of eigenfunctions that physically represent the stable and metastable states of (for example) a hydrogen atom. The modal approach is appropriate; yet it is not easy to follow the process for the system to settle down into the lowest energy state that is most *likely* to happen in an "isolated" atom, or rather in a *statistical ensemble* of such atoms. Correspondingly, in our context and in the context of the Couette motion verified by Taylor's experiments, the process of *evolution* is expected to be even more complicated.

In particular, for galaxy disks evolution depends on dissipative mechanisms in the interstellar medium and on collective effects in the stellar component. This is the main reason for us to focus on the "final" or "current" state of quasi-equilibrium that is likely to have been achieved over a sufficiently long period of time (e.g., a few periods of revolutions) for the *statistical majority* of galaxy disks.

11.5 References

1. Barnes, J.E., and Hernquist, L. 1992, *Ann. Rev. Astron. Astrophys.*, **30**, 705.

2. Bertin, G., Lin, C. C., Lowe, S.A., and Thurstans, R.P. 1989, *Astrophys. J.*, **338**, 104.

3. Butler, K.M., and Farrell, B.F. 1992, *Phys. Fluids*, Ser. A, **4**, 1637.

4. Drazin, P.G., and Reid, W.H. 1981, **Hydrodynamic Stability**, Cambridge University Press, Cambridge.

5. Drury, L.O.C. 1980, *Mon. Not. Roy. Astron. Soc.*, **193**, 337.

6. Farrell, B.F. 1989, *J. Atmos. Sci.*, **46**, 1193.

7. Fung, I. Y.-S. 1977, Ph.D. diss., Massachusetts Institute of Technology.

8. Goldreich, P., and Lynden-Bell, D. 1965, *Mon. Not. Roy. Astron. Soc.*, **130**, 125.

9. Goldreich, P., and Tremaine, S. 1978, *Astrophys. J.*, **222**, 850.

10. Julian, W.H., and Toomre, A. 1966, *Astrophys. J.*, **146**, 810.

11. Landahl, M.T. 1975, *SIAM J. Appl. Math.*, **28**, 735.

12. Lin, C.C. 1955, **The Theory of Hydrodynamic Stability**, Cambridge University Press, Cambridge.

13. Lin, C.C., and Bertin, G. 1995, *Annals N.Y. Acad. Sciences*, **773**, 125.

14. Lin, C.C., and Lowe, S.A. 1990, *Annals N.Y. Acad. Sciences*, **596**, 80.

15. Lindblad, P.O. 1960, *Stockholm Observ. Ann.*, **21**, 3.

16. Mack, L.M. 1976, *J. Fluid Mech.*, **73**, 497.

17. Oort, J.H. 1970, in **The Spiral Structure of Our Galaxy**, IAU Symp. 38, ed. W. Becker and G. Contopoulos, Reidel, Dordrecht, p. 1.

18. Orr, W.McF. 1907, *Proc. R. Ir. Acad.*, Sect. A, **27**, 9 and 69.

19. Rayleigh, Lord 1880, *Proc. London Math. Soc.*, **11**, 57.

20. Salo, H., and Byrd, G. 1994, in **Mass-Transfer Induced Activity in Galaxies**, ed. I. Shlosman, Cambridge University Press, Cambridge, p. 412.

21. Taylor, G.I. 1923, *Phil. Trans. Roy. Soc. London*, Ser. A, **223**, 289.

22. Toomre, A. 1981, in **The Structure and Evolution of Normal Galaxies**, ed. S.M. Fall and D. Lynden-Bell, Cambridge University Press, Cambridge, p. 111.

23. Toomre, A., and Kalnajs, A.J. 1991, in **Dynamics of Disk Galaxies**, ed. B. Sundelius, Goteborg University, Goteborg, p. 341.

24. Trefethen, L N., Trefethen, A.E., Reddy, S.C., and Driscoll, T.A. 1993, *Science*, **261**, 578.

12   A Look into the Future

A long journey has been made since the realm of the nebulae was unveiled by Hubble [2, 3, 4]. The advent of powerful new telescopes and instrumentation, the opening of new observational windows from the ground and from space, and the development of computer technology are all key factors that have allowed us to generate an enormous amount of data and an increasingly detailed view of the complex internal structure of galaxies. Theoretical suggestions were born, and many passed away, as a result of new and improved observations, but we should not forget that theory is often the driving force behind the breakthroughs marked by new observational achievements, even when a given conjecture is proved inadequate by a new set of data.

Theorists usually have an easy time in playing with models—in working out the internal properties of well-defined and selected mathematical entities. The satisfaction can be even greater if the equations that characterize the chosen models can be solved analytically or if they can be formulated in such a way that they are suitable for efficient numerical integration or simulation. This satisfaction reflects the internal consistency of the mathematics of dynamical systems. What is often not addressed in many theoretical articles, or is left to readers to fill in, is the relevance of the model considered to the complex physical system motivating the theoretical investigation. This modeling process generally requires the hardest work and is often a source of frustration. Still, it must be faced if we want to make contact with the real physical world.

In this volume we have adopted a semi-empirical approach in which comparison with the physical facts is given the highest priority. In this regard, some observations, such as the recent infrared images of galaxies (see chapter 5), confirming the general perception in Zwicky's original concept of "coexisting" spiral structures in M51, take on a crucial

significance and are especially gratifying. Thus we have generally focused our attention on the observed morphology of spiral patterns and
on the *maintenance* of the present global spiral structure. The modal
theory, while developed in such a way as to provide interesting and
detailed indications also on the dynamical mechanisms for the *generation* of wave patterns (with the capability of encompassing a variety of
specific physical evolutionary scenarios; see chapter 11), primarily ad-
dresses the observationally relevant context of the current state of the
galaxy, and only to a lesser extent the past processes that led to this
state. Differences in the generation process—the initiation of density
waves and the subsequent development of the pattern—may have relatively little imprint now, after a highly irreversible process of evolution.

Symmetry in nature has always fascinated science [5] and might provide the key that allows the theorist to bridge the gap between mathematics and the physical world. In the context of the realm of the nebulae, the beautiful spiral morphologies have been a great puzzle and
a source of speculation even before people knew that galaxies really
are like island universes, as anticipated by some philosophers of the
past. From a theoretical point of view, the bisymmetry so common and
prominent among spiral galaxies has been sometimes related to the
symmetry breaking in the classical problem of ellipsoidal figures of
equilibrium.

Once we recognize that the presence of symmetry and order in
highly complex systems suggests an intrinsic, simple interpretation in
terms of the laws of physics, progress is achieved by developing quantitative tools that implement some proposed scenarios and sharpen the
dynamical models for the objects considered. Simplicity and symmetry in complex systems are not a surprise specific to the astrophysical
issues we have dealt with. Rather, they are one of the incompletely
resolved and challenging problems of several fields of research that
include hydrodynamics, geophysics, plasma physics, and, in general,
the dynamics of collective systems. We should continue to regard them
as "incompletely resolved problems" to the extent that full nonlinear
theories (see [1]) and experimental confirmations of the kind available
for elementary systems are still missing in this context of collective
dynamics.

In this monograph we have presented a theory for spiral structure
in galaxies that was pioneered by Bertil Lindblad, based on the hypothesis that spiral structure is essentially a cooperative density wave
phenomenon. By now, the theory has developed into a coherent frame-

work to account for a vast number of astrophysical phenomena related to the classification of spiral galaxies and to their large-scale structure. No doubt the theory is still incomplete and possible alternative scenarios (see chapter 11) should be considered and explored in quantitative detail, so as to decide which would have the wider applicability. Such incompleteness, as is the case for other areas of research in science, should not be taken as a failure, but rather as a stimulus to move on to better and more appropriate models. Progress will take place when better observational and theoretical tools are developed.

Some of the areas where progress can be expected in the near future are the following. Theoretical work toward a more thorough and realistic model for a disk of stars and gas will be based on better and more extensive data from all the observational windows available. In particular, the examination of the density distribution along spiral arms in the underlying evolved stellar disk of several galaxies, already begun thanks to the recently acquired imaging capability in the near-infrared (see chapter 5), is going to give important information on the overall gravitational potential and to provide an empirical test on whether regular global spiral structure can indeed be interpreted primarily as a standing wave-pattern. In general, the new data will allow us to to pin down the main observational constraints on individual objects and on categories of spiral galaxies. Thus we should soon be able to develop more advanced models to explain in detail the rich variety of morphologies that are observed, beyond the simple major categories and prototypes addressed so far; more specifically, galaxies with different types of bars, rings, and smooth arms offer a wide choice of interesting problems that appear to be well within reach.

With the rapid development of powerful computers, such improved modeling of the physical properties of the "basic states" of spiral galaxies will open the way to more realistic numerical simulations. Numerical techniques such as n-body simulations are often enjoyed for their ability to display in vivid form the long-term evolution of the models selected for the simulations. The best use for such numerical studies may yet turn out to be their ability to touch on specific nonlinear aspects of the dynamical theory that are currently beyond our reach, such as those related to mode saturation, mode-mode interaction, and the physics of the interstellar medium. These developments offer long-term challenges and will eventually set the foundation for quantitative scenarios of the evolution of spiral galaxies over the cosmological time-scale.

Of course, progess will also uncover a number of surprises. We hope that this monograph also succeeds in setting the proper context and themes for facing these surprises. In the meantime, there is little doubt that the global spiral structure in galaxies can be understood in terms of density waves, with the collective behavior of the stellar system as its foundation, as was first suggested by Bertil Lindblad so many years ago.

12.1 References

1. Heisenberg, W. 1967, *Phys. Today*. **20**, 27.

2. Hubble, E. 1925, *Pub. Amer. Astron. Soc.*, **5**, 261.

3. Hubble, E. 1936, **The Realm of the Nebulae**, Yale University Press, New Haven.

4. Shapley, H. 1943, **Galaxies**, Harvard University Press, Cambridge (3d ed., revised by P.W. Hodge, 1972).

5. Weyl, H. 1952, **Symmetry**, Princeton University Press, Princeton.

Galaxies Illustrated in This Book

NGC	M	Class	Figure
23		Sb	1.13
205		E5	1.2c
221	32	E2	1.2c
224	31	Sb	1.2, 1.13
309		Sc/SBc	5.4 (optical, K′)
521		SBc	plate 5 (K′)
598	33	Sc	1.12, plate 6 (K′)
628	74	Sc	0.2, 4.4 (gas content), 4.8, 4.11
1201		S0	4.12
1232		Sc	3.2, 4.8
1300		SBb	1.14, 5.3 (amplitude modulation)
1357		Sa	5.1
1398		SBb	0.5
1566		Sbc	3.2
1637		Sc	5.5 (optical, K′), plate 7 (K′)
2217		SBa	4.10
2403		Sc	1.8 (rot. curve), 1.9 (phot., rot. curve), 1.12
2841		Sb	0.3
2859		SB0	0.7
2903		Sc	4.4 (gas content)
2997		Sc/SBc	5.6 (optical, K′), plate 8 (K′)
2998		Sc	3.7 (rot. curve)
3031	81	Sb	0.1, 1.6 (radio), 1.11 (UV), 1.13, 3.6 (HI vel. gradients across arms), 4.11, 5.2 (amplitude modulation), 5.9 (HI isovelocity contours)
3115		S0	1.2

NGC	M	Class	Figure
3198		Sc	1.8 (rot. curve), 1.9 (phot., rot. curve), 1.10 (rot. curve), 1.12
3294		Sc	4.8
3351	95	SBb	1.2, 1.13
3433		Sbc	5.1
4254	99	Sc	8.1
4262		SB0	4.10
4314		SBa	1.14, 5.7 (inner spiral)
4321	100	Sc	1.12, 4.8
4394		SBb	4.10
4486	87	E0	1.2, 1.19 (jet)
4535		SBc	5.1
4565		Sb	1.8 (rot. curve)
4594	104	Sa/Sb	1.2, 4.12
4622		Sb	4.11, plate 4 (K')
4725		Sb/SBb	1.13, 4.4 (gas content)
5055	63	Sb	4.12
5194	51	Sc	0.4 (opt.), 0.6 (red), plate 3 (CO), 5.8 (synchr.)
5195		Irr/SB0	0.4 (opt.), 0.6 (red)
5236	83	Sc/SBb	1.14
5364		Sc	1.12, 4.8
5371		Sb	4.4 (gas content)
5394/5395		/Sb	1.13
5457	101	Sc	1.12, 4.8
5907		Sc	1.8 (rot. curve), 2.6 (radio)
6951		SBb/Sb	4.10
7096		Sa	5.1
7331		Sb	4.12
7743		SBa	1.2
Milky Way			1.4 (IR), plate 1 (opt., near and far IR), 6.1 (HI spiral arms), 6.3 (HII regions, spiral arms), 6.4 (minispiral at the Center), 6.5 (HI rot. curve, N vs. S), 6.6 (disk-bulge-halo decomposition of large-scale rot. curve), 6.7 (stellar velocity dispersion profile)
Large Magellanic Cloud		Irr	(1.17)

NGC	M	Class	Figure
UGC			
2259		Sc	1.8 (rot. curve)
2885		Sc	1.8 (rot. curve)
12591		S0/a	1.7 (rot. curve)

Index